Childbirth Educator Manual

© GoMidwife 2015

© 2015 GoMidwife. All rights reserved.
Pontus University Press

GoMidwife is the only training organization to focus on childbirth education as a ministry tool to serve women throughout the nations. Our unique perspective demonstrates both a need to serve women in the developing as well as developed world. All women need educating during their pregnancies. It should not be a luxury reserved for only the few who can afford it.

The education of a man is never completed.

Robert E. Lee

Table of Contents

The Purpose of Childbirth Education

Why Become a Childbirth Educator.. 10

What is the Difference Between an Educator and a Doula............................12

Assisting Mothers Through Pregnancy.. 13

Birth Perspectives.. 14

Setting Expectations.. 15

It is Natural (But it is Not Normal).. 17

Educating the Expectant Father.. 18

What He Needs to Know.. 19

Increase Her Confidence... 22

Promote Original Design.. 24

Facilitate Self-Discovery and Ability.. 25

Shift Paradigms.. 26

Performance Based Birth.. 26

Psychological Preparation... 27

Effects of Childbirth Education... 28

Philosophy of Education

What is a Birth Philosophy... 30

Determine Your Birth Philosophy.. 30

Writing your Birth Philosophy.. 31

Educating the Pregnant Woman

Education is More than Facts.. 33

Learning Styles.. 33

Worldview vs. Technique... 35

Teaching Moms Who Have Had Miscarriages... 35

Teaching Single Moms... 36

Teaching Teen Moms... 37

Teaching Survivors... 37

Birth Networks and Resources... 38

What She Needs to Know

First Trimester

Understanding Conception.. 40

Physiology in Pregnancy.. 42

Helping Women Articulate Their Birth Philosophy................................ 44

Setting Goals and Priorities... 45

Dispelling the Fear.. 47

Fetal Developmental... 49

Fetal Development Explore More.. 49

Anatomy... 50

Anatomy Worksheet... 51

Maternal Development.. 52

Nutritional Needs... 53

Nutrition Diary.. 56

Teratogens... 58

Prenatal Testing... 59

Emotions.. 60

Hormones... 60

Aromatherapy... 61

Second Trimester

Birth Options and Environments... 63

Birth Team.. 64

Birth Setting.. 64

VBAC... 66

Intervention... 67

Cesarean.. 70

Birthing Techniques and Methods... 71

Breathing: Start Now.. 72

Weight in Pregnancy.. 73

Sex and Pregnancy.. 74

Third Trimester

Fetal Kick Counts... 75

Length of Labor.. 76

Physiology in Labor.. 77

Stages of Labor.. 78

Coping Measures 78

Discussing the Birth Plan 82

Signs of Labor 84

Warning Signs 85

Fetal Proteins, Development and the Initiation of Labor 86

Induction 88

Birth Rights 89

Informed Consent 90

Packing Your Birth Bag 91

Preparing Your Home 92

Unmet Expectations 96

Grief and Loss 97

Postpartum and Newborn

Warning Signs for Mother 99

Physiology of Postpartum 100

Expectations 102

The Social Network 102

Safe Sex 103

Family Planning 103

Feminine Hygiene 104

What is Normal and What is Not 106

Warning Signs for Newborn 106

Normal Newborn Development 107

Welcome to Parenthood!.. 108

Breastfeeding.. 110

Community Classes

Centering Pregnancy.. 112

Individual Approach... 112

Adapting to the Needs of the Audience.. 112

Communication Skills

Language.. 113

Culture, Religion and Ethnicity.. 113

Effective Methods.. 114

Teaching Strategies

How to Set Up a Classroom... 118

Classroom Interaction and Participation.. 118

4 Week Class Program... 119

Curriculum Development

Writing Your Own Curriculum.. 122

Creating an Outline.. 124

Writing Lesson Plans... 125

Self Assessment.. 126

Course Capstone... 128

End Notes.. 129

Teach these things to your children, talking about them when you sit at home and when you walk along the road, when you lie down and when you get up.

Deuteronomy 11:19

The Purpose of Childbirth Education

Why Become a Childbirth Educator

In just a few generations we have lost the familiarity and normalcy of birth. When birth was removed from the home, when our mother's packed their bags and left us behind to give birth to our siblings, we lost the connection and understanding of how birth unfolds. The lack of sight led to the poverty of knowledge and understanding, but now we recognize this loss and the detriment of the gap in our intimacy with the design of our bodies. We are aware of the imperative to educate both ourselves and the women around us to grasp the realization of just how to make childbirth normal once more. Our generation, able now more than ever to equip ourselves and others with the knowledge and understanding of childbirth, must be the one to restore the wisdom of the birth process and, until birth is routinely returned to the home, we must do so through classes and education. We must begin to educate both ourselves and the women around us to a greater awareness in how to make childbirth normal once more. Education rests with us and we must educate both in word and deed.

Education can eradicate misconceptions, bring knowledge and displace fear and uncertainty. Throughout society, in every generation, education has at times been withheld from the common people in order to maintain ignorance and push through agendas not necessarily in the people's best interest. Childbirth is no exception. Knowledge is power, and understanding changes outcomes. In educating women and men about pregnancy and birth, light is brought to a subject often shrouded in vagueness and fallacy. Childbirth, like sex, is often not openly discussed, and when it is it is generally only by those who had a poor experience due to a misunderstanding of the process and beauty of the design. Therefore, our only education as women can often be by those who themselves have been misinformed.

Childbirth education is paramount in changing the negative pattern of childbirth we are seeing world-wide. This is true regardless of the circumstances or the location. In the United States cesarean rates are moving toward a drastic 33%:

> Recent studies reaffirm earlier World Health Organization recommendations about optimal rates of cesarean section. The best outcomes for women and babies appear to occur with cesarean section rates of 5% to 10%. Rates above 15% seem to do more harm than good (Althabe and Belizan 2006). The national U.S. cesarean section rate was 4.5% and near this optimal range in 1965 when it was first measured (Taffel et al. 1987). Since then, large groups of healthy, low-risk American women who have received care that enhanced their bodies' innate capacity for giving birth have achieved 4% to 6% cesarean section rates and good overall birth outcomes (Johnson and Daviss 2005, Stapleton et al. 2013). However, the national cesarean section rate is much higher. After steeply increasing over more than a decade, it leveled off at 32.8% in 2010 and 2011 (Hamilton et al. 2012). So, about one mother in three now gives birth by cesarean section.[1]

And in the developing world education could change the approach to family planning and thus reduce the mortality rate of women in more rural nations:

> Women in developing countries have on average many more pregnancies than women in developed countries, and their lifetime risk of death due to pregnancy is higher. A woman's lifetime risk of maternal death – the probability that a 15 year old woman will eventually die from a maternal cause – is 1 in 3700 in developed countries, versus 1 in 160 in developing countries.[2]

We must educate ourselves and then we must begin to do what our grandmothers did; we must live out the understanding that birth is a natural and normal part of life. We must teach what we know; talking about it as we walk with our friends and children, when we eat with them, before we go to bed and when we get up. We must come to know truth in childbirth and that truth must be always on our lips.

The purpose of becoming a childbirth educator is therefore to open the dialogue, expel the myths, and transform the understanding of possibilities surrounding childbirth, one pregnancy at a time. In the West our purpose is to help women sincerely know birth is designed to work, as God intended. In the developing world, where they do still give birth at home, the purpose must be to educate women not how to give birth, but how to have healthy pregnancies so death does not have to so commonly be the outcome. We must then encourage each woman, each community, to become their own educators, seeking out truth for themselves not only in this process but also as they continue to raise their children. It should be the goal of every childbirth educator to equip each mother and each couple with the tools they need to make decisions that will change their families, communities and subsequent generations.

What is the Difference Between an Educator and a Doula

Often a childbirth educator is also a doula or vice-versus, but not always. Commonly, becoming a doula and childbirth educator are some of the first steps to becoming a midwife. Both offer distinctly separate and exceedingly valuable roles.

Although doulas do educate the women they serve, they more commonly do not enter into a woman's life and pregnancy until about the third trimester. This is somewhere around 32 or 36 weeks of gestation, long after many important decisions have been made...or not made. Doulas will then accompany the mother through the last weeks of pregnancy, the labor and birth and into the postpartum weeks as a companion and advocate. The main difference then between a doula and a childbirth educator is the childbirth educator equips a mother throughout her entire pregnancy and the doula assists her through the birth and postpartum period. The partnership then of educator and doula is fundamental to the culmination of a properly prepared woman, and a successful and fulfilling childbearing outcome.

Childbirth educators frequently begin walking alongside women in the earlier weeks of pregnancy, especially with first time moms, as they seek to learn how to care for themselves and their growing baby. The job of the educator is usually complete before labor begins. Educators fill the information and resource role. They equip the mother with the knowledge she needs to be fully prepared to embrace, not only the labor, but all aspects of her pregnancy and postpartum and be well throughout. The role of a childbirth educator includes, but is not limited to the provision of nutritional information, helping the mother and her partner set expectations and providing them each with tools to have a successful pregnancy, labor, birth and postpartum period. Educators enable the woman and prepare her with every resource she needs to find success in each chapter of her birth story.

> *By wisdom a house is built, and through understanding it is established; through knowledge its rooms are filled with rare and beautiful treasures.*
>
> Proverbs 24:3-4

Assisting Mothers Through Pregnancy

ed·u·ca·tion

1. the act or process of imparting or acquiring general knowledge, developing the powers of reasoning and judgment, and generally of preparing oneself or others

2. the act or process of imparting or acquiring particular knowledge or skills

As a childbirth educator (CBE), the main goal is to assist and equip each mother through her pregnancy with knowledge and a better understanding of herself and her body. This is done by using varied avenues of resource and information to impart to her the tools she needs for this particular pregnancy. The best education is not the repeating of facts, but rather the awakening of one's mind to desire to know more. As an educator your job is to help the women, and her husband, to recognize the need for an active role in their own pregnancy and birth. Optimally, you will use diverse methods of teaching including many and varied resources for each couple to engage through their chosen form of mode and media. You should include, but not be limited to: lecture, discussion, pictures, hands-on activities, movies, books and storytelling from other mothers.

The benefit of childbirth education is multifaceted and includes not only providing accurate information to each couple about how to make informed choices throughout this meaningful process, but also filling in the gaping wholes of expectations and dispelling the myths surrounding childbirth. Often, women will assume their education lies in the hands of their healthcare provider, but rarely is that the case making childbirth education imperative. Through education, their confidence in their own ability increases and anxiety, which often surrounds the mysteries of childbearing, is relieved. This then contributes to the overall health and well-being of the mother throughout her entire pregnancy.

> *Whether you think you can or you think you can't, you are right.*
> Henry Ford

Birth Perspectives

Perspective changes everything. How birth is perceived is often how birth will be. Unhealthy perspectives need to, and can, shift during pregnancy. Likewise, long before labor beneficial perspectives should be encouraged and supported. When beginning a class, take time to listen to each person's perspective on labor and birth. Do they come to you with no perspective or expectation? Is natural birth something they believe in already? Or have they brought scars either literal or figurative from past birth experiences? How you approach each class should be determined by the audience and their perspective of birth.

A perspective is an attitude or formed thoughts regarding a particular subject. If we do not actively form our own thoughts about childbirth through self-education, then often our perspective will be formed for us, and that through the narrow lens of our society. The mark of a good educator is not that their students come away with the very same opinion as they themselves hold; in fact the educator's own opinion should not so often be made known. A good educator will present all the perspectives, facts and information and allow the student to arrive at and form their own opinion. If this step is missed, if the educator simply dictates to the class one perspective or a single point of view, then the mothers and fathers might never take hold of a belief for themselves.

Explore More:

Type the following key words into Google and view the images: Nelson Mandela, sculpture, perspective What does this sculpture tell us about perspective? How can this apply to birth and birth education? Using this understanding consider how it relates to birth. How can perspective be changed or challenged? As educators, how do we keep ourselves from being nearsighted?

Setting Expectations

How do we encourage a class who either has no formed thought or perspective on birth, or perhaps they have a negative one, to set expectations of success, beauty, joy and triumph in birth? To understand this more clearly, let's go back and look at the beginning of this course: How were we originally designed to know truth? Remember the verse this course opened with? Talk with them when you're at home, when you walk, before you go to bed and when you get up. In other words, we learn through story.

To set expectations we must hear the stories of others. Think about it: why is there such a fear pervading birth these days? Because those are the stories being told. We have all heard the bad, the ugly, the scary over and over again, but to reform perspective the story must be looked at it from another angle. Think back to the assignment you just completed, looking at the sculpture from one angle, or even from an up-close perspective. You could not see the whole, but step back and the picture was made clear. The same is true when helping a mother or a class to set expectations. They need to be shown the whole picture, not just one perspective. That means as an educator, you must not give one-sided opinions...even if it is the one side you believe, but rather provide for them a broad stroke of information and allow them to decide for themselves.

To help mothers set realistic expectations for their own births they need to hear countless stories in all shapes, sizes and colors. What that means is they need to hear the perfect stories, the stories where there was struggle but success, the stories where there was indeed failure, but rebound with the next pregnancy and birth. They need to hear enough stories where they can begin to connect themselves inside the story itself. Too often mothers are so disconnected from childbirth because it is no longer an everyday normal occurrence we see and in which we play a part; they neither prepare nor set expectations.

What then are realistic expectations? How do we equip women to truly succeed in childbirth? It must be in the stories. It often seems the stories mothers bring to birth are on one extreme pole or the other. Either they say, "my mother/sister/friend had a 6 hour first time birth from start to finish and then sneezed and the baby fell out" or "my mother/sister/friend labored for 3 days, had to have Pitocin, had to have an epidural and could not have the baby naturally and had to have a c-section or they would have both died". These stories do occur, but they are extreme on both counts. Neither of these should be the expectation we as educators or friends allow a mother to take in to her birth. Both polar opposite stories can set her up to have a poor perspective on what birth is truly like. How then do we help a mother set realistic expectations? What should be taught? Every mother should be taught to expect her body to work as it was designed by a God who intricately knit us together, and every mother should be taught her story will be an original. As an educator, the stories you present to the class, be it through movies, books, in-person storytelling (which will connect with more people world-wide) or videos, should cover both sides of the spectrum as well as everything in-between. It is your job to provide such a broad stroke of information, to pull the perspective out wide enough, so they can find their place and see where they fit inside the story. If you do this, they will be able to set realistic expectations that can be met with success.

 **Explore More:**

Seek out a minimum of four mothers and record with audio or video their birth stories to be used as teaching materials for your future CBE classes. These stories should be, on average, three-five minutes long and each should be filmed separately for a total of four individual videos. Remember: do not gravitate to just the perfect stories include the gritty ones too, the ones that highlight strength and achievement. As you hear stories about birth, or work alongside women in the upcoming years, ask to record their stories. Continue to do so and you will have an incredible resource to share with the women you teach.

We don't see things as they are, we see them as we are.

Anaïs Nin

nor·mal

conforming to the standard or the common type; usual; not abnormal; regular; natural.

It is Natural (But it is Not Normal)

Birth is a natural process. However, it is not a normal process in most parts of the world, specifically in the developed world. It is not assumed a woman will labor and give birth without intervention. Natural is not the standard of normal. Why is it necessary to hold classes to educate women on the normalcy of birth? It is because they no longer see the process. Birth is a learned behavior just like walking and talking, both of these processes are natural, but they must be taught and they must be learned.

Expectation, therefore, is the pathway to see a return to normal, and education is a large component of that process. Why do you think most children learn to walk and talk when they do? It is because we expect them to and we work diligently with them until we see them rise up and meet the potential we know is within them. As a good parent, one would never say, 'Walking is natural so you'll figure it out." No, absolutely not. We know it is natural, but we also know for our children to learn, we must walk alongside them and teach them how to successfully navigate the process. It is the same with birth, and you will also find with breastfeeding, it is a natural action that requires development. As we educate ourselves and the women around us, and as our paradigm begins to shift on what birth can and should be, the natural outflow will be to change society and ensuing generations.

 **Explore More:**

Use your browser to search for these key words: "Ted Talk", "Ina May" and watch the Ted Talk video.

© GoMidwife 2015

Ina May is a woman who rose to fame beginning in the early 1960's simply by learning how to trust birth, talking about it and living it out alongside other women. She has, as one woman, changed an entire generation of understanding about birth. She travels and teaches extensively to women, nurses, midwives and doctors alike.

Educating the Expectant Father

When educating the father it is our job to also equip them with realistic expectations for themselves in childbirth and for their own success. They are not doulas nor midwives and we should not expect them to fill a role not intended for them. If the expectation is for them to function as such, it will be the rare individual able to meet those expectations and the rare man who can also own the birth experience. What should be the role of the man in the birth relationship? Let's look at how they were designed and see if we can draw from that.

Men were designed to fill the following roles: protector-defender, provider and leader-teacher. Of course, these are not their only roles, but they are the primary ones which characterize most men. How do we equip them to fulfill these roles in birth?

One of the responsibilities of the birth educator is to support the father in his innate role within the birth process. Birth is a time when he should step forward more fully into the role he was created for, not step back. Birth, like no other time, is when a woman should be protected, provided for and lead. They must not relinquish their authority, nor their place, in the birth room. A father who is convinced of the way childbirth should unfold is the best advocate for the woman and their child.

He will be intimately invested in decisions as the birth unfolds, but he will be removed enough to make logical and clear judgments as needed. He should step forward to protect and provide for her both physically and emotionally. He will know how she desires to see birth take place and as she works hard in labor he will advocate for her needs as the mediator between her and the care providers. Men who step forward in protecting her space and hopes find birth as fulfilling as the woman does. It is his right to guide throughout the birth process, and he should not relinquish this authority. He should be her covering at all times.

Not all relationships will initially be built this way, in fact many will not, but as you encourage and equip him through this process he will be enabled to rise up to his potential alongside her. Birth is not a spectator sport, and men should be as equally invested in the outcome as the woman. Birth can both define and refine us all.

What He Needs to Know

Fathers need to know they are included and have a specific role in pregnancy and childbirth. The woman did not become pregnant alone; in that, the father had a significant part and should not be excluded during the rest of the process. Often, as childbirth providers, we overlook the aspect of the father in pregnancy, labor and birth. We either ignore them entirely or send mixed messages. We do both the mother and the father a disservice when we focus solely on the woman and her role. As a childbirth educator your job is to address the needs of both. Fathers are neither helpless nor incompetent when it comes to childbirth and when equipped for success appropriately. They are in fact half of the childbearing process and bring a strength and objectivity to birth not found in the birth professionals nor the mother. In order to be properly prepared, here is what he needs to know:

Fears- Men have fears too. They are real and more common than we might think. Generally though, these fears can quickly be eliminated. As women, our fears are often based on emotions, how we feel, and that is not so easily rectified. Men, however need facts. Their fears surrounding childbirth are usually due to the lack of knowledge or understanding of the process and subject matter. Once facts are presented and lack of education is dispelled, men readily grasp the truths surrounding birth and can be a stunning defender of both the mother and the natural process. Addressing the fears men have is usually a quicker process as an educator than addressing the fears of the woman.

Role-Some men will want to have a hands-on support role and others will not. Both are okay responses. Either way, they have a role in labor and birth and as their educators it is important to prepare them to fulfill, and exceed, their own expectations.

In Early Labor- It is important for fathers to know the signs of labor...what is real and what is uterine preparation. The loss of the **mucous plug** can be the on-set of labor, or it could mean labor is still days away; usually the later. It is an important role for the father to remain objective during these early signs, because it is often very difficult for the woman to do so. Knowing the mucous plug to be one of the earlier signs, the father can encourage her that her body is beginning to change and remind her it is working correctly, but he can also help keep perspective and choose rest in preparation for the hours and days ahead. When it comes to **Braxton Hicks contractions** versus progressive contractions, it is often hard to tell the difference when you are the one feeling the pain. The pain of Braxton Hicks contractions is a very real pain, and one women do not want to think of as "not active labor". Understanding the physiology of these contractions, however, can help expectant mothers not put too much effort into these types of contractions, even though they might want to. Braxton Hicks contractions are not consistent, do not form a pattern and are irregular in nature. They generally will go away with position change and more often than not are considered and uncomfortable rather than unbearable or progressively painful. True labor contractions usually are described as menstrual cramps and will likely also wrap around from the front through to the back. Being able to help her decipher Braxton Hicks and true labor contractions from one another can make a huge difference in the decision process of when to call the midwife or when to go to the hospital. Once the on-set of true labor has been determined, then knowing how to **correctly time contractions** is a key part of the process. There are several apps which are readily available to help time contractions, but understanding **411** can make all the difference in choosing when to take the next action steps. As much as possible, the father can be the full support leading up to this time. Choosing to labor together for as long as possible, and equipping him with the understanding of how to comfort her during this process, is one of the best ways to encourage labor progression. When contractions are four minutes apart, from the beginning of one until the beginning of the next, and are lasting at least 1 minute in length and have been consistently so for at least one hour then active labor is considered to have begun. Up until this point the objective is to continue the daily routine, rest as much as possible, breathe, relax and work together.

In Active Labor- It is important for the father to also remember to breathe. Around this time, it is quite easy for a new and expectant father to want to hit the panic button, but understanding the process of birth can help him keep his head about him during this time. At this point, whatever birth path you have chosen should be initiated, be it the arrival of the midwife, or the decision to move from the home to the hospital for the birth. Reassuring words and a calm presence are the most important gifts he has to offer during this time. He is often her safe place when nothing else seems to be stable. If he wants to be a hands-on support, then counter pressure usually comes into play here. If she has chosen a water birth, often the father is also in the tub and helping her to maintain position. During this time he can remind her, in a quiet voice, to relax and breathe, just as practiced in class. The woman must now give her entire attention to each contraction and from this point on he will become very much the advocate he has been educated to be.

Advocate- As the advocate he can voice their desires for certain interventions, or their desires to avoid certain interventions. He should be well-versed in the birth plan and can help direct the course from an objective, yet invested position. Having gone through childbirth education himself, he should know the benefits and risks of each intervention offered and can help her navigate the course now that they are on it. One of the key roles a father takes in childbirth is that of the liaison between the mother's expectations and plans, and the healthcare provider's.

The purpose then, of educating both the mother and the father, is to lay the foundation on which the birth team can be built. Birth in this day and age is viewed inside out, with the doctor being the most important figure in the birth and the significance trickling downward to the mother and father somewhere near the bottom of the hierarchy. This should not be the case. Education, then, can revolutionize the understanding of who really matters in labor and birth, the rights a mother has, and can decidedly change the course and outcome of a birth experience.

Notes:_____
_____
_____
_____
_____
_____
_____
_____
_____

With realization of one's own potential & self confidence in one's ability, one can build a better world.

Dalai Lama

Increase Her Confidence

Words are powerful. If you want to know what a women truly believes about childbirth and her ability to give birth, ask her what she says to herself when she is alone. Does she repeat phrases such as, "I am strong enough" and "My body was made for this"? Or does do these words sound more familiar: " I don't know if I can do this" and " I hope I can handle the pain"? Whether words are spoken out or internalized, they will often determine our course. How do we increase the confidence of the women we serve?

No Comparisons

Encourage each women not to compare her body or her birth with any other woman or any other birth, even her own previous ones. Anyone who works in birth will tell you that every story is different and every one unique. Discuss with her how to embrace the unknown, because unknown does not need to equate to fear. In fact, the unknown should be the harbinger of excitement and anticipation. It is an untraveled adventure. She has yet to write her story, and it can be written in the most beautiful way.

Planning

Taking the time to actually plan out how she sees her birth taking place will also increase her confidence. Encourage her to think through her birth from beginning to end. Help her formulate questions to ask herself and suggest she write a birth plan. Where does she plan to give birth? Does her idea of what birth should be match with where she has chosen to give birth? If she has chosen a hospital birth, what is her plan for pain relief? What about laboring and birthing positions? If she has chosen a home or birth center birth, what will be the plan should the need to transport arise? Thinking through these and many other issues will allow her to relax into birth knowing all the details have been thought through and prepared. If you are serving women in a more rural setting, planning is even more imperative. Knowing eventualities have been thought through and planned for can present the woman with the ability to relax and surrender to the birth process. This act of surrender can often make a marked difference in labor and birth.

Setting up a Support Team

It is imperative the birth team have the same values and ideals as she herself holds. This will increase confidence, not only in herself, but also in the process. This team will help enable her to reach the aspirations she has for birth. Provide a resource of doulas, midwives and mother/baby friendly doctors in her area.

Trust in Her Ability

You can embolden her to trust in herself. Many times we just need to know we are believed in and reminded we can give birth without fear. Again, it is the wider perspective that allows us to see the whole where they cannot. Many women who seek out childbirth educators are first-time moms. They do not yet know they can do this, and they have not seen anyone else do it, either. Sometimes, they just need to hear it again and again. Birth professionals, educators, doulas, and midwives, all can inspire women to deepen belief in themselves. It is all a part of our job, and in truth, one of the most rewarding aspects.

Notes:_____
_____
_____
_____
_____
_____
_____
_____
_____

It isn't that they can't see the solution. It is that they can't see the problem.

G.K. Chesterton

Promote Original Design

Birth was designed; it is sophisticated and elaborately planned. Birth and it's process is not an accident and it is not flawed. We must begin to believe this if nations are to be changed through birth. When people intervene, complications increase. This is not to say all hands off as pregnancy and birth do need support, but it should be the proper support of the design, not of a shamelessly flawed man-made system. As you educate women through the childbearing years the significance of the inherent ability within her to give birth must be stated and restated. It is vital she learns about and understands how wondrously she was made. If she can grasp this, she will begin to trust in herself and and her capacity to give birth.

The man-made system is too arrogant to see the problems caused by intervention. It will not be reversed. It will not change, but we can seek to change ourselves and change women one labor and one birth at a time. The organizations worldwide have not been able to stem the consequence of a broken system, and birth as a system is broken. However, we carry with us a piece they do not have. Consider the plan of God for the redemption of the Earth. When He sent the King of Kings to be bring salvation, how did He choose to do it? Salvation was conceived and brought forth through the womb of a woman. In our day, when the Earth is groaning, consider how a loving Father might once again bring forth the salvation so desperately needed through the unexpected and humble wombs of women. He conceives His greatest treasures in hidden places. It is this belief we must grasp and not let go. Each woman and each baby within her womb gives the opportunity for Christ Jesus to be revealed. We must labor to bring birth back to its original intent: to point to the Father and for His glory.

Our goal in education and childbirth work is not merely to change the techniques or reform the system but rather restore the eyes of the people to see their Creator at work in them as they labor and bring forth. It is to reshape the understanding of birth in such a way that God can be made known through the process.

If you must take the lead, lead so that the mother is helped, yet still free and in charge. When the baby is born, the mother will rightly say, "We did it ourselves!"

Lao Tzu

Facilitate Self Discovery and Ability

Awakening her mind to new ideas can lead to self discovery and self discovery will lead to confidence in her own ability. The quote by Lao Tzu is foundational to childbirth education and subsequent prenatal or midwifery care. The goal of every good childbirth class is to facilitate learning and self-discovery in such a way the mother and father will be able to say "they did it themselves". Do not tell a mother what she should believe or how she should give birth. Present the options and let her discover her own ability.

How do you facilitate self-discovery? To begin, you must believe in both her and her ability to give birth as God intended. Before you can help her, you must believe in her; this will open her up to believe in herself. Encourage her to explore her own nature and beliefs about childbirth. Behaviors are learned, and childbirth is no exception. Discuss with each mother what birth experiences they have had and how each one has affected them and their beliefs in their own abilities. Pregnancy is a wonderful time to discover and heal from past events in life. Help her to close the door on any negative past experiences and embrace a new possibility.

Notes:_____
_____
_____
_____
_____
_____
_____
_____
_____
_____

"Do I bring to the moment of birth and not give delivery?" says the Lord.

Isaiah 66:9

Shift Paradigms

As educators, women, mothers and world changers we must begin to facilitate a shift in our paradigm. The standard must be reshaped and the pattern remolded. The reformation of what is acceptable in the sphere of birth begins with radical young women unwilling to compromise the promises of God, for both herself and those she is called to serve. It takes a conscious effort to know what we believe and why we believe it and then to actually walk this belief out. In this way we will begin to shape not only the birth experiences of the women we are serving but the communities themselves. Together, both in word and deed, we will help reshape the worldviews of those not only in need of better birth outcomes but also those that have had little or no voice in choosing what is best for themselves and their families.

Performance Based Birth

Birth requires active participation but is not defined by accomplishment. It is likewise not performance based but is instead a compelling and momentous event in the lives of women. It is in these moments that surround the birth of our children that our very lives pivot. It is mandatory for a woman to truly grasp that she is designed, fashioned, formed, equipped and adequate enough to give birth. Birth is not just something we do. It is not only a verb. Birth is also a noun *and is defined as* **becoming alive**. Our identity does not hinge on where we gave birth or whether or not we did so with or without pain medication or surgery. Of course, we want to encourage women in such a way that they strive for the most excellent, but we do not want do so at the expense of those who cannot, or do not, achieve that goal. There should be no guilt by association to a cesarean. Is it what women aspire to? No. Is it what women should acquiesce to? Absolutely not! Birth settings and birth outcomes in no way negate the validation of motherhood. Your classes should stress this aspect.[3]

Psychological Preparation

Childbirth education can help to prepare a woman not only physically, but psychologically too. Often women will prepare the nursery, eat right, exercise and in all other manner prepare physically for labor and birth...right up until the moment it begins. Then, they will commonly assume it will just all "work" and leave the most important part to chance and in the hands of their care providers. They have prepared for the physical needs, but not the emotional needs. Here, they are completely caught off guard. Most childbirth education classes do not help a woman prepare psychologically for the birth. They discuss comfort measure and positions, which is good, but not enough. This lack of mental preparation can lead to so many needless interventions. A woman must *believe* she can have a baby and do it successfully. The psychological preparedness of a mother is more key to a successful labor and birth than many choose to acknowledge.

Here's how a mother can prepare psychologically for birth:

Choose a care provider with the same values and beliefs in birth, someone she is comfortable with and who has taken the time to know her and her story.

Choose a support team

Understand the birth process: What does NORMAL birth actually look like

Understand, but not dwell on possible complications. Determine the appropriate response in the event of their occurrence.

Think through scenarios and determine the emotions that might arise: Contractions begin the day after the date she is due to give birth. They come on strong, but then stop. This happens for several days with no "real" labor beginning. How does that make her feel? What are the emotions? (Surrender is a part of this scenario too!)

Listen to positive stories.

Stop listening to negative ones.

Think positive: about herself, birth, labor, the baby

Trust her ability and know her strength

Surrender She cannot make this happen. Doctors and midwives cannot make this happen. If she has prepared physically, now also mentally and emotionally, then when labor begin she can truly sink into it and relax knowing all eventualities have been prepared for. Surrender leads to relaxation. Relaxation alleviates tension and without tension pain is lessened.

Effects of Childbirth Education

Childbirth education impacts childbirth outcomes. A recent study reported women who took childbirth classes were more likely to:

- Communicate better with the fathers of the baby
- Relax during labor
- Feel confident and prepared
- Be more for postpartum and breastfeeding
- Improved health
- Better self care and awareness
- Improved perceptions of labor and birth[4]

Each class should promote discussion and discovery. Many women are verbal processors, and need a safe place to talk through thoughts and ideas, especially if the ideas are relatively new to them. Communication is key to achieve our purposes in childbirth education.

 **Enrichment:**

List a few potential discussion questions that you have thought of while working through this material. Consider making a running list of interesting topics to add interest to coming classes:

1. _____
2. _____
3. _____
4. _____
5. _____

Philosophy of Education

phi·los·o·phy

a particular system of thought

Developing your philosophy of birth is foundational to how you will teach and educate the women you serve. Do not be surprised if your philosophy changes over time, but you have to begin somewhere. If you have given birth then stop and consider how you came to the conclusions you did. Did you philosophy change after you gave birth or was is the same after as it was before? If you have not given birth, then you will approach the development of philosophy very differently. You need to know what you believe about birth and develop your own personal philosophy. This is important for several reasons, your philosophy will influences your teaching and it is fundamental you articulate your philosophy of birth to the mothers and women you teach. Your philosophy will infuse every choice you make from the words you use, the resources you recommend and the care providers you promote.

It is vital, as you set your philosophy, to be honest with yourself. If you are not honest with how you feel about birth personally you will not be able to evoke the same trust in birth from the women you serve. Identify the emotions that surround birth for you and then build your philosophy from the ground up.

What is a Birth Philosophy

A birth philosophy is your particular system of thought regarding birth, what it is and how it should be approached . Do you trust in the body's ability to birth or do you think medical intervention is necessary? Do you believe birth is better at home or must it be in a hospital? What do you think is the most important aspect of pregnancy and what do you feel influences good outcomes more than any other thing? How should a woman approach labor? Should she always have a support team or is that optional? Each aspect from conception to postpartum should be considered.

What you believe is what you will teach. You cannot teach something a part from the belief system you hold. Your philosophy on pregnancy and birth will permeate every aspect of your education and class.

Determine Your Birth Philosophy

Ask yourself the following questions:

What words do I associate with birth?
When I think about birth I am...excited, nervous, pensive, afraid, trusting, anxious....
What do I believe about the birth process? It is a perfect design. It is safe. It is holy. It is scary. It is messy. It works...sometimes. Or some other thoughts.
What do I believe about the labor and birth process? It never requires intervention. It sometimes requires intervention. It will almost always require some kind of intervention.
Labor is...normal, dangerous, painful, safe, scary, a natural process.
Birth requires...patience, prayer, a doula, a midwife, no one but the mother, always a doctor.
Birth should take place...in the home, in a birth center, in a facility with a surgical center and a top NICU center.

Labor and birth should...never require pain medicine, should sometimes require pain medicine, should never be felt, numb from the neck down please.

Take time to consider these questions and more. Be honest with yourself and the answers you give. It is important to do so, because as educators we hold a great deal of influence over the lives of women. Take time to be quiet before the Lord and ask Him what His views on birth are, He most certainly is not silent on the subject. In fact, you will find the Lord uses birth analogy quite often. Once you have determined the essential, core words to describe your pregnancy and birth values begin to write your philosophy statement.

Writing your Birth Philosophy Statement

When writing your philosophy it can be as long or as short as you like. Make sure it is concise and clear. Perhaps you will need to write a longer statement at first then take from that original detailed philosophy and condense the content down to the most important points of your beliefs.

A personal example: I have written many philosophy statements, some very long, but in the end my beliefs hinge on three things: 1. I believe birth is designed by God and is a very spiritual matter. In fact, I believe birth is the story of the Gospels. Birth can be used to witness to every woman around the world and we must get back to the original design so the picture of Jesus can shine through birth. 2. I believe the foundational building block to good healthy pregnancies is nutrition and 3. I believe women need women to help them navigate pregnancy, labor and birth. I could write a book based on my philosophy, but here it is condensed down to the most important aspects so when I present my philosophy to a class or an individual it is concise and clear, and class you will know immediately who I am and what I am going to teach. This is my birth philosophy:

Women are strong, and birth is originated by God; perfectly and intricately designed. Birth illustrates the Gospel and deliberately paints the picture of Christ as it unfolds unhindered. This beautiful and complex design should be bolstered through nutrition. Most of the common problems which occur in pregnancy and birth can be prevented through proper nutrition. Labor should be supported and the woman should be supported by those who love Jesus, love her and love birth. Complications do arise; birth should be attended at home when possible, by educated observers with the capacity to respond in an emergency situation. The baby is the mother's and should go immediately to her chest and remain.

"It is not only that we want to bring about an easy labor, without risking injury to the mother or the child; we must go further. We must understand that childbirth is fundamentally a spiritual, as well as a physical, achievement. The birth of a child is the ultimate perfection of human love." Dr. Grantly Dick-Read, 1953

Enrichment:

Take time to consider, formulate and write your birth philosophy. This should be articulated in such a manner that you will be able to give it to any individual or class you teach in the future. Length is not the desired outcome, simplicity, honesty and your values are. Consider being fairly generic in your religious views, if you mention them at all. I state mine religious beliefs clearly in my philosophy because I know my audience and we are like-minded. If you plan to work in various parts of the developing world and among certain cultures and people groups this should be taken in to consideration as you write your statement. This statement may change as you continue your journey as a birth educator, but be sure to carefully consider what you truly believe about childbirth. It will be very important, when a woman is choosing her educator, for there to be no doubts the platform from which you will teach.

Notes:_____
_____
_____
_____
_____
_____
_____
_____
_____
_____
_____
_____
_____
_____

Educating the Pregnant Woman

Education is More than Facts

Some childbirth classes are the mere recitation of facts, but when you are using education as a ministry to build relationship and change lives and outcomes then education must be much more than the simple sharing of retained data. Education becomes the impartation of knowledge, skills and understanding. The end goal should not be the woman successfully took your birthing class, but rather she successfully gave birth. Your responsibility to each woman does not end at the last class, it ends after she has birthed successfully. Education is not the handing off of a few books but, according to the Wikipedia definition, education is a general sense of learning in which knowledge and skills are transferred from one generation to the next through teaching and training. It goes on to say that any experience that has a formative effect on the way one thinks, feels, or acts is an education. Each class should form or reform the birth paradigm.

Learning Styles

Learning styles are the individual way in which a person acquires, processes and retains information and knowledge. People vary in the way they learn, and in order to be an effective teacher you much incorporate all styles of learning.

There are three main learning styles:

- Auditory
- Visual
- Kinesthetic

An auditory learner is one who learns through hearing. They will retain information when you speak to them. Auditory learners learn best through lectures and the reading and reciting of material out loud, including textbooks.[5] A way of incorporating auditory learning during childbirth classes is through discussion questions.

A visual learner needs to see what you are teaching. They learn best when graphs and pictures are used as well as movies. They can learn through lecture if they can see the lecturer and visuals are used. Visual learners can retain things better when they write it down. Because visual learners are just that, visual, then they are easily distracted should something else come into their line of vision. Flashcards can help them study and worksheets can be included in class or as homework to help them take ownership of what they have learned.

Kinesthetic learners need to learn through hands-on materials. They need to experience the subject. Kinesthetic learners need to be doing something as they learn such as standing or as part of a group. They learn well through demonstration, classroom activities and fieldwork.

There are other learning styles as well, and they include:

Aural which is a form of auditory but uses sound, especially music.[6]

Verbal, which prefers using words specifically speech and writing

Logical learns using logic, reasoning and systems.

Social learners prefer to learn in groups or with other people.

Solitary learners prefer to work alone and use self-study.

A good educator will include all of the major learning styles throughout each class in order to be able to teach well to all audiences.

 **Explore More:**

world·view -The overall perspective from which one sees and interprets the world. A collection of beliefs about life and the universe held by an individual or a group

Worldview vs. Technique

One's worldview is based on what they believe to be true. We all have a worldview and it is shaped primarily by the community in which we live. The mother can know all the right techniques and all the positions for birth, but if her worldview is scanted then no amount of technique will prevail over ingrained beliefs. The beliefs must be changed, and that through education. Worldview can change, and often does, but not without new information that counters old belief systems. A mother's worldview of the life she carries within her, of who she is as a woman, of how her society sees and shapes birth determines what she thinks is possible. This returns us back to the psychological preparation: what she thinks possible, or impossible, will influence the outcomes of pregnancy and birth. To change birth, systematically, we must change worldviews.

Explore More: What is an example of a worldview that would change a birth outcome? How would you counter the worldview through education?

Teaching Moms Who Have Had Miscarriages

When working with couples who have gone through a miscarriage and are now expecting again, you will likely need to address grief and loss along with worry and fear that the loss may occur again. Anxiety will often invade even the most healthy of pregnancies during some point, but for mothers who have had a loss this feeling is even more paralyzing. Not addressing the concerns would be an oversight, but addressing them in a routine class would not be not ideal. Suggest a time where you can meet with this couple one on one and can give them your undivided attention. Often pregnant couples, particularly mothers, who have experienced loss do not feel they are understood by their healthcare providers nor is their loss or grief recognized. All too often we do not know what to say, so we say nothing at all, and her loss is overlooked; she is left to grieve alone. Miscarriage is a common occurrence in women of childbearing years, but it does not make the individualized experience any less traumatic. We must, therefore, care for each mother on an individual basis. Simply listening to her story and acknowledging her loss is often enough. Grief is natural and grief is a process. Do not be afraid of her grief. However, if the grief, fear or anxiety is significant and effecting the current pregnancy, be equipped with resources and counselors to help her process these emotions in a deeper way.

Teaching Single Moms

First, one of the most important roles you have as an educator working with single moms is to help them find and build a support team. Birth is is very much a team sport. It takes many hands to help a woman through labor and even more words of encouragement; a woman should never choose to enter birth alone.

> It takes a thousand praise to birth a baby!
>
> Chinese Proverb

It is important too, to recognize they may be grieving. The sudden loss of relationship can cause the already volatile emotions of pregnancy to be intensified. Help her grieve, but also help her recognize the importance of celebration. New life is a joy and this needs to be forefront when possible.

Single moms are brave and strong; help her to celebrate that bravery and not be ashamed. Shame is not a normal emotion of pregnancy, but in particular situations it too can come in to play. Regardless, it has no business in the pregnancy or birth room. Encourage them to bring along a doula or family member to support them and be aware and mindful not to use exclusive terminology in class which would cause them to feel isolated. There is no reason a single mom cannot be as successful in her pregnancy and labor as anyone else, but it is our job to make sure she has the additional support and resources she may need to make this an amazing time for her.

Teaching Teen Moms

There are several issues facing teen moms. Besides commonly being single moms, teenagers often have diets that provide little proper nutrition and usually have no understanding of their bodies, anatomy or the process of pregnancy and birth. Most women who are outside of their teenage years have some of these concepts and have at least read about pregnancy as they generally are planning their pregnancies. When educating teen moms there are two approaches: 1. hold a class with only teen moms so you can start from the foundation and really search and talk through many unexplored topics or 2. you can have an integrated class so the teenagers can also learn from the older women in the group. Childbirth education for teenage mothers is going to take more work and much more follow up than other clients will require. It must be understood going in that they may not absorb or be willing to follow the advice you set forth. Nonetheless, there will be the ones who do and it is for them we do our job and do it with joy.

Teaching Survivors

Teaching survivors is one of the most sacred and potentially redeeming roles we can aspire to as a childbirth educator. We must approach them gently yet understand survivors are strong women, possibly stronger they we are ourselves, and this is their moment. Birth can be healing. Where violation and death have betrayed their bodies, now life has the occasion to recapture their beauty and breathe a new essence over them. Where ruin and devastation have been, life can now come forth. Katie Wise, herself a survivor, says the following are key aspects necessary for survivors who plan to give birth:

Permission: I will be choosing to allow this baby to spread my pelvic bones wide as I welcome him into my arms.

Love: This baby was created from an act of love, as is giving birth.

Protection: The people around me, as opposed to my perpetrator, are there to protect and support me.

Power: I will give birth. I will actively work with my baby to create a miracle. Very different, indeed.

If you plan to teach survivors, be encouraged to read *When Survivors Give Birth* by Penny Simpkin. As with teenagers, they will require more time and effort. There will likely be a higher level of processing needed and the usual fears and anxieties are going to be heightened. You are there in both regards not only as an instructor and educator, but also as a facilitator and encourager. You are now part of their birth team and will help to bring healing in a new and unique way.[7]

Birth Networks and Resources

Enrichment: Approximately 1/3 of what you provide families will be in the formal setting of a classroom. That means the rest of the education process will happen outside of the classroom and most likely as they explore the resources you provide. Take time to develop your birth network and the resources you have available for your mothers and fathers. Research and create a comprehensive list of reliable resources for your mothers. Include names, phone numbers and an address, when possible. You will need to know not only who is in your area, but also their philosophy of care. For example, which pediatricians will delay vaccines and which ones will not, or which hospitals allow VBAC. For resources such as books, some CBE's choose to put together a lending library and share these resources with their clients. Even if you do not choose to do so, be sure you have a ready list of books, movies and websites for them to explore. This is a list that will be ever growing and changing as your own network and resource list develops. As you compose your list, think about what each of the following have to offer a pregnant woman. What are the various perspectives of care and what type of care might be the norm with each provider or place? When choosing movies or books think about what the agenda is for each resource you use. As an educator you do not want to present just one-sided resources. You must educate and allow the woman to make her own choices. It can be very easy to only present one view of birth, and although we want to promote natural birth and perhaps even home birth, we must also present all options to be fair and balanced.

Midwives (CPM, CNM):_____

OBGYNs_____

Hospitals/Birth Centers_____

Pediatrician_____

Naturopaths_____

Chiropractors_____

Massage Therapists_____

Nutritionists_____

Lactation Consultants_____

Doulas_____

Breastfeeding Support Groups_____

Mommy and Me Groups_____

Dinner Doulas _____

Books_____

Movies_____

Websites_____

Other_____

Prenatal care is actually what occurs in-between each visit to your care provider,

not what happens during the 15-30 minutes you spend with them..
Amy Kirbow

What She Needs to Know

What does a pregnant woman need to know in order to best care for herself and the growing life inside? As an educator you must be thorough and teach each mother about childbirth beginning at conception and continuing through until the extended postpartum period. It is right to assume you may be her only resource for what she will learn and know about pregnancy, childbirth and her newborn. With this assumption in mind, educate each mother that prenatal care truly is the care she gives herself and her baby all those weeks in between visits to her care provider.

First Trimester

Understanding Conception

To many, conception is a complicated mystery unable to be understood. Indeed it is complicated, but it is also one of the most amazing and beautiful processes in the human body and points so clearly to One who knit us together. Most women, especially teenagers, do not know the complete process of conception. This does not need to be a science class, but a thorough understanding of what has occurred or what will occur can help many women understand better how to care for themselves and prepare for pregnancy in the future. If the women do understand their bodies well and the cycle of fertilization and conception then often the husbands and boyfriends will not. In the developing world it is almost always the case that they have not been taught. Pictures and movies are the best way to introduce this subject. Use a movie or pictures and then have an open discussion answering questions they may still have.

Keep in mind, when choosing pictures, who the audience is. In oral learning societies they do not conceptualize as we might.

If, for example, you show them a picture such as this[8]:

to demonstrate fertilization, they might see the sperm literally and will then believe the sperm is the length of your hand, when in actuality they are microscopic and this picture is increased hundreds of times in order to be visible at all.

 **Enrichment:**

As you begin this assignment consider who your audience will be. Are you going to Afghanistan? Then write this as if you were teaching Afghani women. What do they need to know? How do they learn best? Do you plan to teach teenage moms? How will they relate to conception? How might they best be engaged and learn?

Choose:
1. A ten slide PowerPoint presentation
2. A ten page storyboard
3. Other Medium

Clearly create a presentation depicting conception. Your presentation should be between 4-5 minutes in length. Be sure it is engaging, clear, colorful and accurate.

Any change, even a change for the better, is always accompanied by drawbacks and discomforts.
Arnold Bennett

Physiology in Pregnancy

Physiology is the study of the function of a living system. In pregnancy physiology looks at how a woman's body changes and adapts to accommodate the growing life within her. The physiology of pregnancy is complex yet flawlessly designed and takes place right alongside her growing baby. Nearly every organ will make a physiological adjustment during pregnancy. Physiological changes include not only the transformation to her physical appearance, but also changes occurring deep inside the body. Pregnancy especially effects the endocrine, renal, cardiac, respiratory, and hematologic systems. The purpose of the physiology is to protect, support and prepare both the mother and the baby for birth. When a woman is educated on the the normal changes to expect during pregnancy she is able to identify when physiological changes move outside of normal. The mother's age and genetics will play a large part in her physiological response, but proper nutrition can often make up for any deficit.

Endocrine: the endocrine system is responsible for the production of hormones. Without the proper physiological changes taking place in the endocrine system the pregnancy cannot be sustained and labor, birth and lactation cannot take place. Many of these necessary changes are also what cause discomforts for mom. An example would be the production of the hormone relaxin. Although necessary to relax the pelvis in order for baby to pass through, it also relaxes the esophageal muscles leading to acid reflux and relaxes the veins leading to varicose veins.

Renal: the kidneys are quite taxed during pregnancy. In the early months of pregnancy renal flow is increased by approximately 40% and the kidneys themselves increase in size 1 cm. Urine production is increased, and thanks to the endocrine system working properly, progesterone effects the urinary tract by relaxing the smooth muscles surrounding it, which can often lead to UTI's or urinary tract infections.

Respiratory: the lungs change both structurally and functionally to adapt to the demands of mother and baby. As the metabolic system adapts and increases it places an heightened demand for additional oxygen on the respiratory system. In early pregnancy a mother will often feel a shortness of breath due to this. Later in pregnancy shortness of breath is more likely to be related to compression from the ever growing baby.

Hematologic: the requirement during pregnancy on the components of the blood are profound. The blood is responsible for transporting oxygen from the lungs to the tissues and carbon dioxide back to the lungs. Thus an obvious increase in the respiratory system effects the blood. Blood volume increases dramatically in pregnancy, up to 45%, to accommodate such needs. Not only does blood transport, but is also responsible for the body's defense system and is also heightened to protect both the mother and baby. This system must not only rise to meet the immediate needs, but also begin to increase physiologically in order to recover after birth through increased clotting and iron stores. These normal physiologic changes can sometimes lead to abnormal complication such as thrombosis.

Cardiovascular: physiologic changes occur in the cardiovascular system and include anatomical changes. The heart increases in size during pregnancy to accommodate the growing demands put on it by fetal development. The maternal heart should return to size just after birth. Beginning in the second trimester the maternal pulse will increase approximately 15 beats per minute. As the pulse increases, it is common to see blood pressure decrease due to the expanding blood volume in the second trimester with it rising back to pregnancy rates in the third trimester. It is also not uncommon for palpitation or benign murmurs to be detected during pregnancy because of the system's adaptations.

While these systems are greatly effected by pregnancy, they are not alone. There is not one system to remain physiologically untouched. Massive changes are also taking place in the metabolic, integumentary, neurological, skeletal, muscular, and obviously the reproductive systems including the uterus and cervix. While normal physiology must take place in order for growth to occur, it is important to support these changing systems with proper nutrition for well-being[9].

Enrichment:

The physiological changes in pregnancy are generally what cause common discomforts. Some normal discomforts include: pelvic pain, heartburn, acid reflux, varicose veins, fatigue, nausea, constipation and insomnia. Research the common discomforts of pregnancy and create a handout with natural remedies for each complaint including foods, exercises and herbs.

Helping Women Articulate Their Birth Philosophy

Now that you have written your birth philosophy you will better know how to help the women in your class to write their own birth philosophy. Just as it was important for you as the educator to write down your thoughts and ideas of birth, it is equally important for her, the consumer, to do so as well. Birth is a significant passage in the life of a woman, and all too often they approach the momentous chapter in their story without too much reflection of forethought. This one misstep alone can cause poor outcomes or, through the process of education, can help lead women along the way of discovery triumph within their own stories.

A birth philosophy is different than the birth plan. The philosophy of birth is the idea, theory, or way we think about a subject, ours being birth. The philosophy is what we believe to be true, and the birth plan is how to use various methods and measures to accomplish and achieve said philosophy. This philosophy statement should reflect her beliefs about labor and birth and is an exercise in self-discovery and articulation. These should be generalized ideas that consider her values, attitude and beliefs regarding the birthing process. All good philosophy statements must come from the heart and herald the core beliefs of a woman and/or the father of the baby. A good exercise might be to have her write her philosophy at the beginning of the class and then another one at the end to see if any of her beliefs have been influenced or changed due to education.

Setting Goals and Priorities

Setting goals and priorities are of utmost importance when it comes to planning for birth. Goals should not be set arbitrarily, and the means is not the goal but rather the end. As you determine your goals and set your priorities, do so by embracing the beauty of the goal itself. Since birth is so organic, compromise may be a real factor as birth does not always go as planned, even in the best and most glorious circumstances. It is necessary to predetermine where lines will be drawn and concessions will not be an option. A woman's identity is very much woven together within her birth story and can often define her, or at least mark her, for a lifetime. Birth is a significant and irreversible event and must be approached with reverence and deliberation. It is true that a woman will labor in her mind long before her body surrenders to the physiological process. So much of the pain women endure in labor can be negated with clear, precise information as well as both mental and physical preparation. In order to set goals and priorities it is essential to spend time in self-reflection and decide what is imperative and which beliefs are non-negotiable.

Begin by having her answer the following questions:

-**What are the five most important aspects about labor and birth to her?** Think deeply here and not just superficially. Yes, it is important to be able to move during labor, but go a little farther and answer based on your core beliefs about birth. Think of the things that are most important to you about this time and the process. Encourage her to think about the indelible imprint which will remain behind on her heart. How will she FEEL during birth, how does she want to feel, and what will lend to that accomplishment.

-**What birth stories have influenced her and why?** She is writing her own story and no story will ever be the same, nevertheless she can take inspiration from others who have successfully gone before her. What story does she aspire to? Which stories inspire her? Be sure to urge her not to absorb the negative stories which are bound to arise as she seeks out the stories of others.

-**What ideals does she live by already that can be incorporated into her birth?** Too often we leave our beliefs at the front door when we arrive at the hospital. Standards we would never agree to in our everyday life are accepted during labor and birth. This should not be so. For example, having others make decisions for you. How often do you go to a restaurant and the waiter tells you what you will or will not eat? This does not routinely happen in life at large, but it does in birth.

As a woman sits down to begin to write encourage her to write an initial statement which clarifies who she is and then states any significant events in her life which have lead her to her conclusion and beliefs about birth.

Creating and writing a birth philosophy will take more time than the class will allow for. This can be a homework assignment you give to each couple prior to their arrival in class where you have them form the framework at home and then you can allot 5 minutes to write it out after the class has discussed the importance of a birth philosophy, or you can have them bring it pre-written philosophies and use those willing to share as a discussion point.

Notes:_____
_____
_____
_____
_____
_____
_____
_____
_____
_____
_____
_____
_____
_____

I just say to myself, "I know you are afraid, do it anyway.."

Unknown

Dispelling the Fear

At one point or another during pregnancy fears are likely to arise. Whether it is the fear of pain, the fear of being out of control, the fear of losing the baby or the fear of induction and cesarean, fear is real. Not only is fear real, but it is intense and can be overwhelming. Part of the fear arises out of the unknown, and once a mother receives education and answers then some of those fears will be alleviated, but other fears arise from the download of the fight or flight hormone. Fears are normal...it is how one responds to those fears that will make all the difference. It is critical to recognize each fear, validate her feelings, talk through each and every one and then help her to walk in the opposite spirit. Dispelling fear prior to labor is one of the fundamental proactive steps that can be done to prepare for labor and birth. Every good educator, doula and midwife knows fear breeds pain. If you can eliminate fear then you will likely achieve the birth you desire. Laboring to bring about the birth you dream of prior to the physical aspect of labor, through dispelling fear, should always be one of your main goals, so be on the lookout for opportunities to do exactly this. Encourage mothers to dispel the fear through:

Education- This is obviously your greatest tool. You are an educator. Fear can be dispelled when knowledge is at the forefront. Once you learn where their fears lie, you can tailor classes to speak to those fears and neutralize them. Sometimes a mother will just need to verbally process her fears and talk them through. Be ready to listen, but always advocate for her to process and let them go. Fear can be a chapter in the story, but it should never remain the focus. Again, you are a resource; should a mother need more than you can provide, be sure to connect her to the proper counselor or professional.

Preparation- Knowing you are prepared for a big event can allow you to relax, and birth is not an exception. If a woman is prepared nutritionally, physically and emotionally then she can rest when the time arises and allow her body to do most of the work. Encourage her to prepare her body with proper protein and hydration as well as through exercises and relaxation methods. Give her every tool she needs to be prepared and she will excel.

Join the Conversation- Advocate for your mothers to not be silent nor afraid to ask questions. Provide them with resources of positive birth experiences or connect them with women who have positive birth stories. Positive stories do not necessarily equate to natural home birth births, but they include any story where the mother feels inspired by her own birth and is wiling to share. Perhaps as part of the education process have a story telling night. Invite back mothers from previous classes to come share with moms-to-be.

Listen- Listen to her and teach her to listen to her body. We are a society who is generally so busy we do not take time to listen to our bodies, but if we would then we would find they are intelligent and work remarkably well with few exceptions. Advise her to begin to listen to her body early in pregnancy and practice throughout gaining a better intuition and the ability to hear her bodies urges and respond to them appropriately. One of the first ways she can begin to listen to her body, if she is free from chemical addictions such as tobacco, caffeine or alcohol, is to respond to the cravings her body has for nutrition and various foods. Whether it is the foods we eat, the activities we do, or any other aspect of how our bodies function it is important to take the time to simply pay attention to what it might be saying to us.

Believe- Finally, believe in her and teach her to believe in herself, her intuition and her body. We are not taught to trust ourselves, especially in medicine. Instead, we are taught to trust the doctor, but doctors canNOT birth this baby. Only she can do that, and if she will trust herself she will know the best way to do so. This belief can be a powerful tool for her to push aside fear and let her body do what it was designed to do. In some case it is enough for her to be reminded that she can do this and, if she truly believes it, she can finally release her tension and fear. In times such as these the actual birth may not be far behind.

© GoMidwife 2015

Fetal Development

It is important for each mother to know how her baby should be developing. There are clear markers, even in the womb, which signify proper growth and development. Recognizing these milestones can assure mothers and caregivers alike that development is progressing as it should. This is not an embryology course, so in-depth details of development are not necessary. However, certain stages of development can be supported by certain foods. For example, when the brain is at the height of development a mother should eat foods high in good fats and omega-three fatty acids. When the skeletal system is beginning to solidify then calcium should be on the menu. Also, it is comforting to know when she should begin to feel the baby move and how big the baby is during that particular week. This knowledge can help a mother begin to bond with her baby long before she will hold baby in her arms.

 **Explore More:**

Often in developing communities you will not only teach the women but also the children, and if it is interesting enough the men might join too. In our society the men and older children are almost always a part of the child birth classes. Keep this in mind and consider your audience. Are you teaching in a rural community in Nepal, an urban setting in the heart of Detroit or young couples in London? Think outside of the proverbial box as you consider how to teach fetal development. Often when we sit down to create lesson plans we do so in a lecture based format, but teaching should be varied and engaging. There are few limits to teaching consider the following: art, song, dance, felt, puppets, or a homemade movie. How do you effectively demonstrate and actively teach a subject if language is a barrier? How do you engage the participants in this topic?

Female Anatomy

Surprisingly, most women do not know their own anatomy. Understanding their anatomy is fundamental to understanding birth.

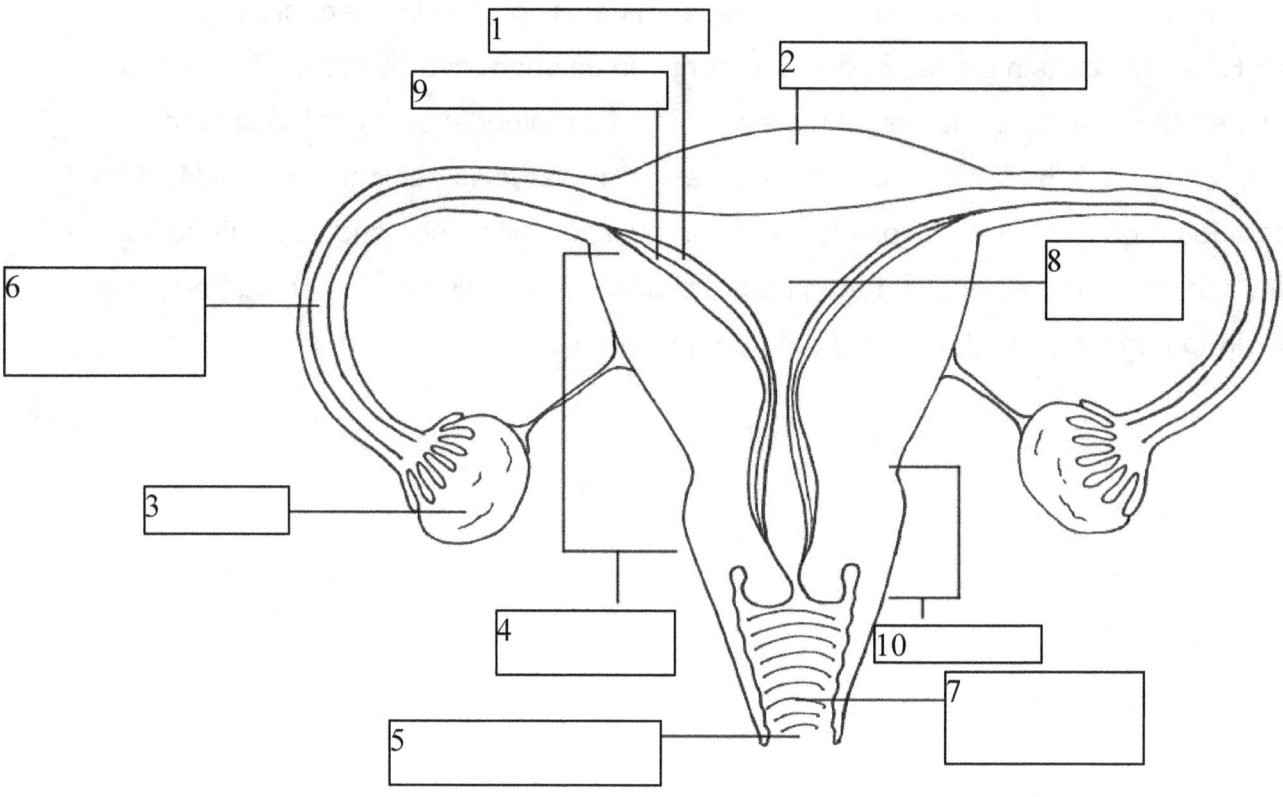

© Copyright 2015, GoMidwife

1. _____
2. _____
3. _____
4. _____
5. _____
6. _____

5. _____
6. _____
7. _____
8. _____
9. _____
10. _____

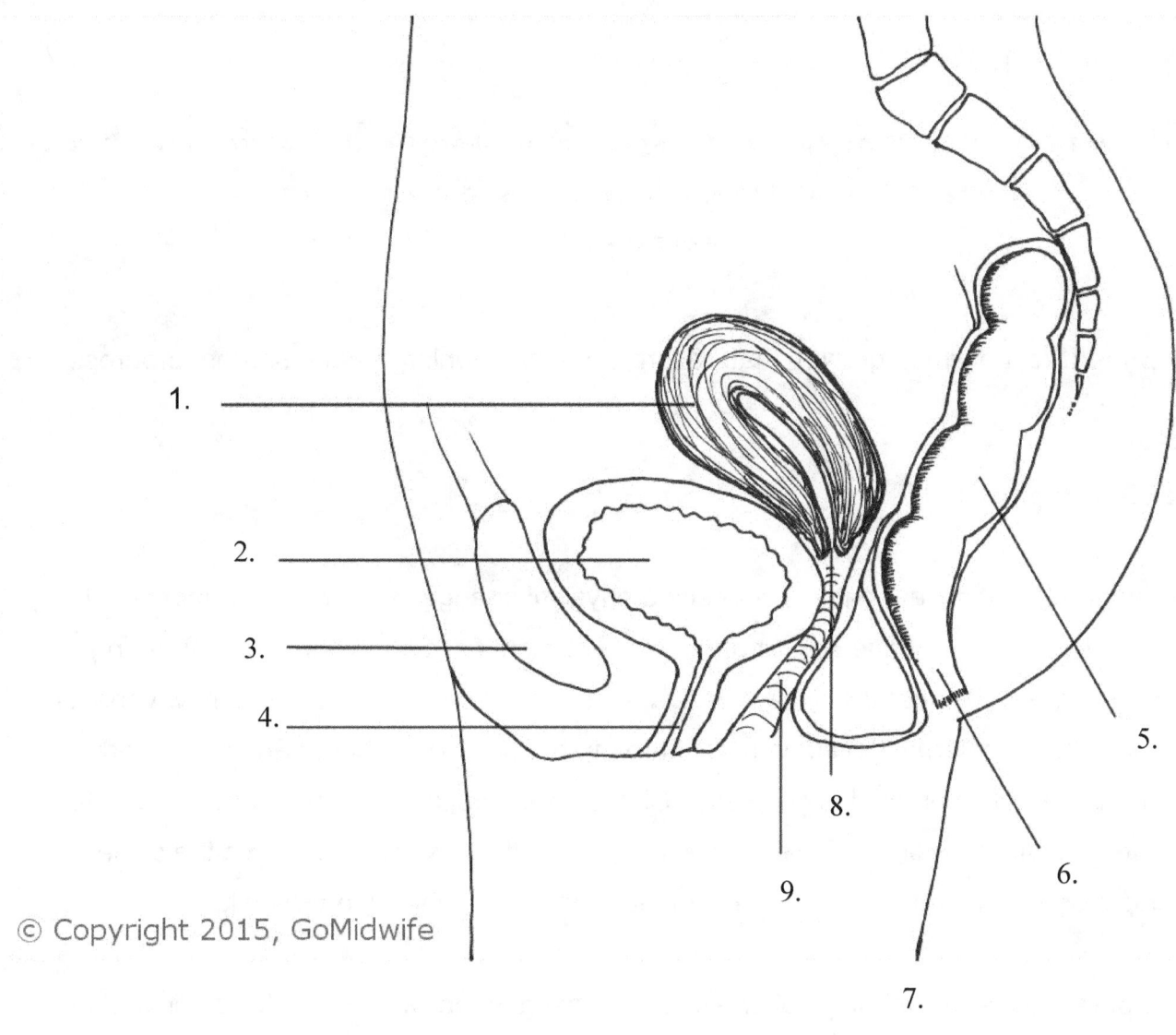

1. _____
2. _____
3. _____
4. _____
5. _____
6. _____
7. _____
8. _____
9. _____

The ability of a woman to alter herself to support and nurture the development of another being within her own body is one of nature's most impressive feats.
Unknown

Instead of wishing away nine months of pregnancy, I'd have cherished every moment and realized that the wonderment growing inside me was the only chance in life to assist God in a miracle.
Erma Bombeck

de·vel·op·ment (synonyms) growth, maturation, expansion, enlargement, spread, progress

Maternal Development

It should be noted that once a baby is conceived physical changes will occur in a mother's body regardless of whether or not she participates. Development from a woman into a nurturing mother however, is quite another. How she adapts, copes and grows with these new variations of normal is, in fact, maternal development. If the mother can be educated and supported properly then she will embrace this chrysalis. Childbirth education is a small part of a greater whole. If we educate the woman, we educate the society. If the woman develops the society will develop along with her for the woman is the foundation on which we are built.

There are physiological, psychological and spiritual changes that will occur. We are in fact a body, soul and spirit and no one part remains untouched by pregnancy. Some changes she will see, others she will not. The experience of pregnancy allows for not only the growth of the baby and an expanding belly, but also for her growth as a woman. Much of the development a women will experience occurs during the pregnancy process. As some would say, our very purpose is fulfilled when we grow life within our wombs. Pregnancy is a unique occurrence not granted to all. It is in that intimacy and struggle to embrace the change that causes women to plumb their depths and it is often here we come to know more of ourselves.
We tend to fight against the transformations of pregnancy, wishing to remain as we were before: physically, emotionally and hormonally. It is the embracing of that struggle, however, where we can find the most freedom.

© GoMidwife 2015

When you know better - you do better.

Maya Angelou

Nutritional Needs

Nutrition is the foundation of wellness. On average a woman should gain between 20-30 lbs while pregnant. If she comes into the pregnancy underweight then she can gain more and if overweight she can gain less. A pregnant woman should consume between 2,000-2,500 intentionally, nourishing calories per day. The growing baby requires both micro- and macro-nutrients to grow and thrive, and almost all pregnancy related complications and diseases can be prevented if a mother is properly nourished throughout her pregnancy. Proper nourishment allows us, and our growing baby, to reach full potential. A good healthy diet will consist of a variety of whole and colorful foods. What are whole foods? Whole foods are ones as close to their original state as possible, with little or no man-refined processing. What foods do you need to be properly nourished during pregnancy and while breastfeeding? You will need foods full of the essential nutrients your body needs to build a strong and healthy baby such as proteins, green leafy vegetables, fruits, whole grains, seeds and nuts and dairy.

Let's begin with one of the most abundant substances in our body and the building block of our cells: **protein**. Women should consume at least 100 grams (**100 grams= 400 calories daily**) of protein per day while pregnant. Proteins can be found in seafood, lean red meat, poultry, eggs, beans, tofu, milk, cheese, yogurt, nuts and seeds.

Pregnant women can safely consume 12 ounces of seafood a week per week, according to the FDA and fish that are safe to consume are: salmon (high in omega- 3 fatty acids and critical for brain development), sardines, anchovies and trout. Seeds and nuts of choice include: pumpkins, watermelon and sunflower as well as peanuts and almonds. The body does not store protein (unused protein is stored as fat) so a consistent consumption is necessary.

Carbohydrates equaling at least 250 grams (250 grams=1,000 daily calories). This gives our body energy and provides our brain, muscle tissue and nervous system with vital means to function. Carbs are generally the largest part of our diet and can be stored as fat for later energy use. It is vital to consume proper amounts of carbs, if not the body will use proteins and fats for energy, but they are needed for a different job. You will need to consume both simple carbohydrates consisting of berries, apples, green peas, sweet potatoes and complex carbohydrates such as beans, bran, potatoes, brown rice and whole grains.

Fats equaling about 77-80 grams (77 grams= 700 calories daily) are calorie packed and provide energy when your carb stores are exhausted. Fats are needed in our diet for many reasons, including helping the body to uptake fat soluble vitamins such as A,D,E and K. You can find healthy fat sources in flax seed oil, olive oil, coconut oil, avocados, seeds, nuts, nut butters, butter, and salmon.

Extra calories should be consumed mainly from proteins or complex carbohydrates.

Now let's look at the most important micro-nutrients for healthy development. Micro nutrients include all the vitamins and minerals our bodies need for sustained health including vitamin A, B-Complex, C, D, E and K. All of these vitamins can be found in green leafy vegetable, colorful vegetables, fruits and herbs. Consuming a variety of colorful fruits and vegetables daily will give you most of the vitamins you will need. Vitamin B1 is the only vitamin not found abundantly in fruits and veg although it can be found in kale. It can be found in yeast, liver, pork and cereal grains which have been fortified. **Folic Acid 800mcg** is another vitamin we need to be mindful to take in sufficient amounts. It is water soluble and part of the B vitamin complex responsible for red blood cell formation and brain development in your baby. Insufficient amounts of this vitamin can lead to neural tube defects in early fetal development. (**Iron 27mg daily**) is chief among the minerals needed. Its main function is red blood cell production. It is a blood builder which will bring oxygen to your developing baby and prevent disease as well as aid in muscle contractions and stress resistance. Iron also prevents postpartum hemorrhage. Healthy sources of iron include dark green leafy vegetables, dried fruits such as raisins, figs, apricots, dates, mango and cherries, dried beans, red meat, eggs, nettles and alfalfa.

Calcium 1000 mg this mineral forms the bones and teeth of the your baby. It also maintains the health of your bones and teeth as well as nervous system and connective tissue. Calcium also aids in blood coagulation and maintaining a regular heart rhythm. Calcium can be found in green leafy vegetables, hard cheeses, yogurt, milk, almonds, nettles, alfalfa, red raspberry leaves, dandelion and kelp.

Fluids in Pregnancy water! water!! water!!! The most abundant substance in our body, water will flush out toxins and help our cells maintain health. You can also add in herbal teas and the occasional fruit juice. Coffee and tea are not bad in moderation, 1-2 cups a day should be plenty, but be mindful to wash down the caffeine with water!!

 **Enrichment:**

1. Create a nutritional diary handout. This handout should both be informative to you the educator and to her the mother. There should be two components, one will be for her to fill out. This should document what she has eaten in a week (or two). It should have places for her to document each meal, snack, liquids and supplements. The second component should be informative to her. This can be approached from various perspectives. One example would be a handout on best food sources for protein, iron etc.

2. Create a nutritional smoothie recipe which is balanced and nourishing. Know why each ingredient is included and what it does in the body to provide nutrients and what body system(s) it will effect in pregnancy. Think about creating a smoothie that is specific to a physiological change or a common discomfort. For example: "The Popeye" for mild iron deficiency, or a smoothie specific to build the elasticity of vein walls, or even a Super Food smoothie for daily micro-nutrients.

The following is an example of a nutritional diary to be used by pregnant women. Often, we think we are eating better than we really are, and when we actually keep track of what we are eating we see that there is way more sugar and way less protein in our daily diets than we thought. Protein is the building block of our cells and as we grow another human it is vital we have essential and complete proteins on board.

Nutritional Diary

	Monday	Tuesday	Wednesday	Thursday	Friday	Saturday	Sunday
Breakfast							
Snack							
Lunch							
Snack							
Dinner							
Snack							
Liquids							
Vitamins							
Supplements							
Exercise							

© GoMidwife 2015

List healthy sources of the following:

Protein	Iron	Magnesium	Vitamin C	Healhy Fats
___	___	___	___	___
___	___	___	___	___
___	___	___	___	___
___	___	___	___	___
___	___	___	___	___
___	___	___	___	___

Potassium	Calcium	Folate	Vitamin B & B 12	Complex Carb.
___	___	___	___	___
___	___	___	___	___
___	___	___	___	___
___	___	___	___	___
___	___	___	___	___
___	___	___	___	___

Notes:_____

_____
_____
_____
_____
_____
_____
_____
_____
_____
_____

© GoMidwife 2015

Teratogens

Teratogens are toxins which when exposed to in pregnancy or while nursing can and do cause a birth and developmental defects. The developing baby is especially sensitive to these toxins during certain periods. Once the mothers blood supply is connected to the fetus, approximately 10-14 days post conception, the risk is said to be at its greatest for teratogen exposure affect. The central nervous system of the developing baby is sensitive to toxins throughout the entire pregnancy which is why alcohol is one of the greatest dangers to development. Others, like chicken pox, are only really concerning between weeks 13-20. There are many hazards we live with daily without much thought, but must be considered in pregnancy.

Common Teratogens

Cigarettes
Alcohol
Street Drugs
OTC Drugs
Microwaved foods
Lunch meats with nitrates and other processing chemicals
Mercury found in larger fish
Proscription Drugs such as antidepressants
Lead
Toxoplasmosis
Arsenic
Paint

List Other Terratogens:

1._____
2._____
3._____
4._____
5._____

Prenatal Testing

What prenatal tests will she be offered during her pregnancy? Will she be offered any at all? In the developing world a mother may see a care provider only one to three times during her entire pregnancy. What are the most important tests and which ones, if any, may be eliminated with informed consent?

Prenatal Panel- this will include a myriad of tests including: chlamydia, gonorrhea, hepatitis B, rubella immunity and HIV. It will also include a CBC. Usually offered at the initial visits.

CBC- this test will look at blood type, hematocrit, hemoglobin and antibodies. Usually offered at the initial visit.

MSAFP, Quad or Triple Screen- the test offered will depend on the lab used by the care provider. This is a genetic test to determine if there is the potential for chromosomal anomalies such as Trisomy 21 and Trisomy 18. Usually offered at 12-16 weeks.

GDM- this tests blood sugar levels and is looking for gestational diabetes markers. This test is usually offered between 24-28 weeks.

H/H- this is a redraw of hemoglobin and hematocrit. Usually at 28 weeks.

GBS- this test is looking for a bacteria commonly occurring in the vagina or the rectum. The test is generally performed between 36-38 weeks.

Part of education is informed consent. Too often in our society medical procedures and lab testing are presumed. The client is not informed nor given the chance to decide for themselves if this test is right for them. Not all tests are necessary for all people. It is very important to encourage your client to research each test and know why the test is being performed. Ultimately, it is her decision, but it needs to be an informed decision.

Emotions

As with fears it is important to validate a woman's emotions. The emotions of pregnancy, and birth, are similar in the cyclical nature as is grief. Often, after addressing a certain emotion a woman is dealing with, she will feel relief and it will not surface again for days, perhaps weeks. Over time, however, they can creep back in and, if not addressed, can become a problem. Again, this is an emotional time with all sorts of feelings that are new and unique, perhaps never having been examined or experienced by the mother before. She may not even be able to explain how or why she is feeling the way she does. It is our job to help her explore these and begin to make sense of them, find an outlet for her feelings, and ensure she is able to process them with a person that cares for her.

 **Enrichment:**

There are countless hormones involved in conception, sustaining the pregnancy, birth and lactation. Below is a list of the most common hormones which effect pregnancy and the ones you will need to know as a childbirth educator. Research the following hormones and explain when they are most prevalent in pregnancy and what their effects are in pregnancy. What are the benefits of the hormones and what are the discomforts a woman may experience because of the increase in hormone levels?

Relaxin_____
_____
_____

Progesterone_____
_____
_____

Estrogen_____
_____
_____

Oxytocin_____
_____
_____

Prostaglandins_____
_____
_____

Adrenalin_____
_____
_____

Endorphins_____
_____
_____

Aromatherapy

Aromatherapy is a wonderful tool for both pregnancy and labor. There is a lot of differing information on essential oils and pregnancy, so as an educator, you will need to strongly advise each woman to do her own due diligence prior to using any oils even when only used for aromatherepeutic influences. Aromatherapy is just that, therapy based on the fragrance or the aroma and is not the consumption of oils internally. However, because oils can be absorbed into the blood stream via inhalation, the same cautions must be considered, even when using oils aromatically. We will discuss the oils generally considered safe in pregnancy when not used topically or internally.

With all that said, Aromatherapy can be very influential on the emotional health in pregnancy. Our sense of smell has long held a strong influence on our emotions and can impact the way we feel physically. It is also well documented that our sense of smell can elicit memories for years to come when that smell is around us. Utilizing wonderful scents during pregnancy can not only effect our mood, but can also create a memory you can return to time and again. Aromatherapy can be utilized with one simple oil or can be blended for more of a synergy encompassing several fragrances.

Anxiety- lavender

Nausea- mint, lemon, ginger, lavender,

Fatigue-lavender, chamomile, frankincense

Fear- bergamot, lavender, geranium, chamomile, frankincense

Stress- lavender, lemon, peppermint

Insomnia- lavender

Labor- clary sage, rose, jasmine

 **Explore More:**

The following are also considered safe in pregnancy when used as aromatherapy. Research possible maladies they could be used for in pregnancy and why they might be recommended.

Grapefruit

Mandarin

Neroli

Patchouli

Petitgrain

Rosewood

Sandalwood

Tangerine

Tea Tree[10]

Second Trimester

Birth Options and Environments

There are options for birth that many women do not know they have. When choosing where or how to give birth the woman, and her family, must consider their own birth philosophy. What do they believe about birth, and which birth environment will best support those beliefs. This is not a small decision. Deciding where to give birth and with whom can be life altering. The memory of birth will last a lifetime. Birth will be one of the most significant events in her life, and how she is made to feel during this time will scarcely fade with time.

Encourage her to ask these questions:

1. What is the most important aspect of birth to me?
   These answers can range from: self-led birth, surgical center, pain medicine, freedom to move or a variety of other answers.
2. Does this particular environment support my beliefs in birth?
3. Will the care provider I see for prenatals be the one to be in the birth room?
4. What is the cost? Is it covered by insurance? Is it worth self-pay to get the birth I most want?
5. How do they make me feel when I am not in labor?
6. Do they take time to answer my questions and get to know me as an individual?
7. What if I do not want to vaccinate right away?
8. What are their C-section rates?
9. What are their transfer rates (if home birth midwife)?
10. What is their philosophy of birth?
11. What do they believe about interventions?
12. What do they believe about postdates?
13. What do they believe about natural birth?
14. What do they believe about epidurals and pain management?

There are a myriad of other questions to ask. Asking these questions will open the mother, and

© GoMidwife 2015

family, to an honest discussion and likely lead to the mother formulating her own questions.

Birth Team

Choosing a birth team is foundational to achieving the desired birth outcome. This team should work tirelessly with the mom to help her realize and set birth goals. This is not always the case. Often, choice is limited due to insurance coverage or location, but it is nevertheless crucial to educate families on the importance of making informed choices, and not merely based on the nearest or cheapest option.

Who should be a part of the birth team? Help her consider what each of the following practitioners would contribute to her birth. What is their perspective on birth and how does she envision them as part of the team? When considering health care practitioners, it is necessary to note whether or not birth will be mother-led, participatory and collaborative or doctor-led.

OB
Nurse
Midwife
Doula
Postpartum Doula

Birth Settings

Various birth settings will yield various labor outcomes. It is important to help her explore her options. Some women will not know home birth is even an option, or that a birth center is nearby. Once they have decided on which setting they prefer, encourage them to seek out the the most experienced and credentialed care provider.

Home – Choosing a home birth will lend to a more mother-led labor and childbirth, but it is important to know who you are hiring as your midwife. Home birth is accompanied by a different set of risks than a woman would face should she choose to birth in the modern tradition of a hospital. What it is important for her to understand is that life is risk. Bringing life into the world has risks no matter where she births. A woman needs only to decide which set of risks are acceptable based on the outcome she desires.

Birth Center- Birth Centers can be the middle ground. A woman choosing a mother-led birth, yet not comfortable in giving birth in her home, might choose a birth center birth. Often freestanding birth centers have doctor back-up and hospital affiliation, but not always. Birth centers tend to be more medical in their endeavors than home birth, yet they usually offer a very relaxed and easy pace for labor and birth. Birth centers will have protocols they have to follow, but generally not as arduous as a hospital. Should a true emergency arise, a transfer would still be necessary.

Hospital- Hospital births are the modern norm. Only 4% of the American population will choose to birth outside a hospital be it at home (1%) or in a birth center (3%). Unfortunately our modern understanding of birth is that it is a disease and hospitals, more often than not, treat it is as such. A woman choosing a hospital birth will need more childbirth education and support than a woman who chooses an alternative birth setting. It is imperative that we, as educators, equip women with knowledge and the understanding of informed consent so they are prepared prior to labor and birth.

Notes:

VBAC- Vaginal Birth After Cesarian

Many women are told a vaginal birth is not an option once they have had a Cesarian, and in truth their options are quite limited. Many doctors, hospitals and even states refuse to allow a woman to have a VBAC, or if they do allow a trial of labor, the boundaries are so limiting another C-section generally ensues. It is necessary a woman who has had a previous C-section understand the risks and know her options.

The risk linked with a C-section are greater than the risks associated with vaginal birth after a c-section, which is only a 1 % risk. The American College of Obstetrics and Gynecology (ACOG) agree VBAC birth is low risk and should ultimately be decided upon by the mother, not the facility.

The following are the guidelines put out by ACOG.

VBAC supporters will welcome the following statements from ACOG's revised guidelines:

VBAC is associated with decreased maternal morbidity and a decreased risk of complications with future pregnancies and births. With a VBAC women can avoid complications of multiple repeat cesareans including infection, blood transfusions, bowel and bladder injury, and placental complications (placenta previa, accreta, and percreta).

- The risk of uterine rupture with one prior low-transverse uterine scar is low, 0.5% to 0.9%.
- About 60 to 80 percent of women who labor after a prior cesarean have a VBAC.
- Most women with one prior cesarean with a low-transverse uterine scar should be counseled about VBAC and offered a trial of labor.
- Women with a twin pregnancy, an anticipated big baby, with two prior cesareans, and women who do not go into labor at term can still plan a VBAC.
- With a breech, women can choose to have an external cephalic version (ECV) after the 37th week, an effective procedure that may turn a breech into a head-down position.
- Care providers should discuss the risks and benefits of VBAC and routine repeat cesarean with their patients early in pregnancy and document it in their medical record.

© GoMidwife 2015

- The ultimate decision to plan a VBAC or to have a routine repeat cesarean should be made by the patient in consultation with her provider.
- Providers or hospitals who cannot or will not provide care for women who want to labor for a VBAC should refer women to VBAC supportive physicians and maternity centers.
- Women can request an epidural for pain relief in labor.[11]

This is an exceptional resource for women desiring a vaginal birth: www.vbac.com

Intervention

in·ter·ven·tion

1. the act or fact of intervening.
2. interference of one state in the affairs of another.

Synonyms for intervention:

- mediation
- interference
- arbitration
- **intercession**
- interposition
- interruption

Intervention is not the enemy; unnecessary intervention is. To intervene is to interfere in someones life; to make decisions for them they did not ask for and do not want. For too long, medical decisions have been made based on "medicine by authority". If you are not a doctor, not been to school for 6+ years, do not have a medical degree then you do not have the authority to make the decision. EVEN IF IT IS YOUR BODY! This has been the long-running mentality and as hospitals and doctors have been ever and more beholden to litigation and insurance policies this medicine by authority has become "medicine by intimidation". The hospitals are intimidated by the threat of litigation should a mistake be made. Doctors are intimidated by hospital policy and women are intimidated by doctors. Intervention is the new normal. It has become the standard. Now, in the developing world the opposite is true. No one intervenes, even when it is necessitated. So, education can bring balance.

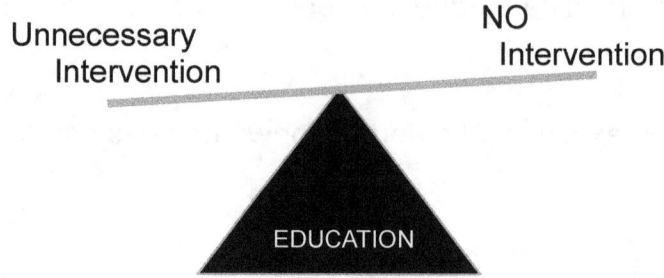

A woman must be the authority on her own body and birth. She alone has the greatest vested interest in her well-being. Although others might be more than happy to make decisions for her, they will never live with the consequences of those decisions; she alone will.

Notes:_____
_____
_____
_____
_____
_____
_____
_____
_____

Before anything else, preparation is the key to success.

Alexander Graham Bell

Being prepared for birth will almost always negate the need for intervention. There are times when intervention will be necessary, but those will be the exception, not the standard.

What interventions are likely (in a hospital) and how can they be avoided.

Routine Fetal Monitoring- this is not necessary an intervention, however it does interfere with normal birth processes so it has been added here. Routine monitoring will necessitate the mother being in the bed and supine. This is not a productive course of action for a natural progressive birth.

Pain Relief- no one likes to see another in pain. So, as the pain increases pain medication will be offered. It is easy to maintain when pain medications are not an option, but once they are available it is hard to say no. Pain medications interfere with the neural pathways of birth.

Cervidil- this is a common drug used to ripen the cervix.

Pitocin- this drug is extremely common in a hospital setting. It will be the rare case to be in a hospital and not receive Pitocin. Pitocin however, is too strong, too consistent for the common woman to be able to endure. This one intervention almost always leads to the epidural.

Epidural- an epidural is one of the most common and most unnecessary drug available in birth. It only become necessary after the normal pain inhibiting hormones are disrupted by any of the above listed interventions. Epidurals are spinal blocks and often limit or complete exclude mobility.

Vacuum- vacuums are often used to assist birth during the second stage where Epidurals have rendered the mother unable to feel the urge to push and cannot get enough power behind the push. In the developing world, where young pelvises are not always mature or malnutrition has caused developmental impairment, then a vacuum can be pivotal and necessary.

Birth is the act or process of bearing and bringing forth offspring.
Wikipedia

Cesarean

A cesarean is an abused and avoidable intervention. "The national U.S. cesarean rate in 2012 was 32.8%."[12] That means 1 in 3 women never "bring forth" their babies. If birth means, to bear and bring forth then it must be stated women are being robbed of actually giving birth. They bear life, but they never bring it forth, rather it is brought forth for them. This is wrong on every level, emotional, physical and spiritual. Women were made to give birth. "To give birth" is an action.

Cesareans are not normal, and should only be used in true emergency situations. They should be avoided if at all possible. Education and preparation can achieve improved outcomes.

 **Explore More:**

Watch the following video. It concerns cesarean sections and is graphic in nature. This video is not recommended for the mother's viewing. As the educator, deeper understanding can lend to more resolve and passion when teaching. Take time to determine your own philosophy on cesareans. What do you believe and why?

https://www.youtube.com/watch?v=r_9TJ6cV128

Method: An established, habitual, logical, or prescribed practice or systematic process of achieving certain ends with accuracy and efficiency[13].

Birthing Techniques and Methods

The main reason many women will attend childbirth classes is to learn the secret of how to give birth. Childbirth educators illuminate the unknown and make the ever elusive natural birth not only knowable, but attainable. There are many birthing techniques available, in the end all techniques lead to the same door: relax, believe, surrender and breathe. If a woman can truly grasp these four concepts and implement them during labor, she will find success.

Explore More: Below are listed the four most commonly known and used birthing methods. Research each method and create an informational handout for each one. Consider how to clearly and concisely introduce and explain each method.

Bradly
Lamaze
Alexander
HypnoBirthing

Often it is a combination of several methods that truly make up the ideal birth approach. Many times women will approach birth with great zeal only to become frightened as labor progresses in intensity and that same energy converts into tension. Methods and techniques are only effective if their implementation can be sustained throughout the entirety of labor and birth, and that when coupled with surrender to the process. A large part of surrender lies in trust. Does she trust herself, does she trust her environment, and does she trust the process. If then, she can fully surrender, technique will often give way to intuition. The best method, or prescribed practice, for most women to achieve this transition is an experienced doula.

Breathing: Start Now

It is important to encourage good breathing techniques before they are required in labor. This is often a good exercise to do in class. Our breathing is connected to our state of mind. When we are anxious or afraid, our breathing becomes more erratic and labored. Our breathing is the most rhythmic and easy when we are sleeping because we are completely relaxed. It takes complete surrender or significant concentration to breathe steadily and rhythmically. Promote relaxed practice breathing so each mother can practice focusing and centering her mind. Have her try her breathing during various parts of the day, and especially when she finds herself dealing with some of those volatile emotions discussed earlier. Proper breathing techniques also provide an increased oxygen flow to the baby, which is an added benefit, and if she knows she is doing something wonderful for her baby it may motivate her to practice more often.

Have each mother start with a cleansing breath: a deep inhale through the nose and a long exhale from the mouth. She can then choose to focus externally, perhaps on someones face, or on an inanimate object. She can even choose to focus on an internal object in her mind's eye such as a beach with waves crashing. Have her slow the pace of her breathing and try to gather a rhythmic cadence in and out, steady and slow. Be sure to encourage her to breathe deep and draw the breath in from deep in her belly. During this exercise it is good if she has someone speaking affirmations quietly to her. The goal is to bring focus and control over the situation, and practicing now will only make things easier to grasp, understand, and work through once labor begins.

Notes:_____
_____
_____
_____
_____
_____
_____
_____
_____

Weight in Pregnancy

In our society, women are often afraid of gaining weight in pregnancy relating all weight gain to fat gain, but it simply is not true. The weight is distributed as follows:

Baby: 7.5 pounds

Placenta: 1.5 pounds

Amniotic fluid: 2-3 pounds

Breast tissue: 2-3 pounds

Blood and Fluid: 4-8 pounds

Stored fat for delivery and breastfeeding: 5-9 pounds

Larger uterus: 2 pounds

Total: 25-35 pounds

When a woman can view the weight to be purposeful then it is easier for her to relax and feed her body and baby well and without concern. In all honesty, if a woman is free of chemical influences, disease and eats whole, non-processed foods, then she can gain weight freely and without too much thought. Independent from disease or excessive sugars a woman will not grow a baby too large for her to birth naturally and vaginally.[14]

Notes_____

_____
_____
_____
_____
_____
_____
_____
_____

Sex and Pregnancy

Some women will have an increased sex drive with pregnancy while others will have a decreased libido. Unless bleeding or contractions are present or sex is otherwise contraindicated by the healthcare provider they are seeing then sex is safe and healthy. As her belly grows, sex as they have routinely known may need to change and other positions explored. Vaginal secretions usually increase as the pregnancy progresses and allows for pleasurable penetration, however should various hormones cause an adverse effect and lead to vaginal dryness then coconut oil is a natural alternative to store-bought lubrication and is also an anti-fungal. It is important to note a baby will neither feel, remember, nor be harmed by the act of sexual intercourse.

Hire a Doula

Although not technically a method, an experienced doula is the most significant birth asset available to women. Choosing to have a doula present at a birth has shown appreciable improvement in a woman's capacity to give birth naturally. Doulas are especially beneficial for "disadvantaged mothers at risk for adverse birth outcomes"[15] such as teenage, low income, or first time moms. A doula should also be considered if the pregnant woman is single. The continual reassuring presence and the emotional, physical an informational support increases the woman's scope of ability. You will want to provide each mother with a resource bank of trusted doulas in her community.

Consider Water Birth

Choosing a water birth, like choosing a doula, is coupled with numerous advantages. All of the above methods and techniques can be actualized alongside a doula and inside a tub. Water birth has yet to be embraced with equality, but the advantages far outweigh any associated risk.
Known Benefits Include:

- Decreased first stage of labor
- Increased comfort in labor
- Increased mobility
- Lowered blood pressure
- Decreased second stage of labor
- Increased relaxation
- Decreased perineal tears
- Increased sense of privacy
- Gentler transition for baby

Resource: www.waterbirth.org

Third Trimester

Fetal Kick Counts

A sure sign of fetal well being is consistent fetal movement and activity. It is possible for a mother to know the routines of her baby inside the womb. One way to maintain assurance of feel well-being is through using the fetal kick count method. All babies will have their own routine. Two mothers sitting side by side will not have the same results. Each baby will have their own rhythm and and move different amounts at different times throughout the day, but all babies will move and move frequently. A mother will likely already know her child's active and sleep periods. The fetal kick count will require her to take note during her baby's most active period. Once each mother has identified the baby's active period encourage her to choose a time each day (preferably the same time each day) to lie down and pay specific attention to her baby's movement. She should note 10 movements within a 2 hour period of time for assurance. Record each day, the time began and the time when 10 movements were achieved. Also, note when each movement was made. Most kick counts are done beginning around 32 weeks of pregnancy, but she can begin sooner.[16]

Week _____   Fetal Kick Count

	M	T	W	Th	F	S	Su
00							
10							
20							
30							
40							
50							
60							
00							
10							
20							
30							
40							
50							
60							

Example:

Week #28

Monday 9:00 XXXXXXXXXX 9:32 Total: 32 min

Tuesday 12:00 XXXXXXXXXX 12:45 Total: 45 min

Wednesday 9:00 XXXXXXXXXX 10:00 Total: 1 hr.

Thursday 9:00 XXXXXXXXXX 11:15 Total: 2 hrs. 15 min.

(This is an example of significant change. In a case like this you should notify your health care provider.)

Length of Labor

A firm fact: the length of labor varies. No two labors will be the same and no length of labor can be determined. However, there are averages for the length of labor. It is always best to plan for the "worst case scenario" when it comes to labor length and if labor happens to be shorter, then the mother can consider it a gift. Not long ago most doctors and midwives based labor on what was called the Friedman's Curve, and although many still do, it is quite important to help a mother understand that as long as there is progress and both she and the baby are well, the the length of labor and its progression does have to follow the curve. It is generally accepted that a first time mom will likely labor longer than with subsequent pregnancies; the second labor usually being the quickest. On average for a first time mom in labor the times are as follows and a second time mom can cut these lengths in half:

First Stage:
Early Labor - 6-12 hours
Active Labor - 4-8 hours

Second Stage:
15 minutes - 3hours

Third Stage:
Approximately 30 minutes

As you can see, labor can last up to 24 hours. It is such an important key for mothers-to-be to understand that when she has the on-set of the first contraction that it is possible, but not likely, she will hold her baby before the same time the next day. This is why rest in the early hours of labor, which is the longest stage, is so crucial.

Physiology in Labor

The physiology of labor is a purposeful set of sequential events with two actions: to dilate the cervix and to push the baby through the birth canal and into the world. These events take place over a period of weeks, days and hours. Both the mother ad baby work together in an integrated process for the culmination of birth. What transpires inside the womb is unseen, but the mechanics and function are known. Fear regularly accompanying the foreign. It is critical then, for each mother to become familiar with the events about to take place inside her body, so knowledge will dispel fear.

Stages of Labor

Labor encompasses three stages. Contained within the first stage of labor are: early, active and transition, the second stage of labor is the pushing period and the third stage of labor is the expulsion of the placenta.

Early Labor: Usually this stage of labor works to thin out or efface the cervix. Some dilation occurs generally from 0-4 cm.

Active Labor: Noted change takes place and contractions become more intense. Active labor is commonly considered to take place from 4 cm-8 cm.

Transition: Begins around 8 cm and lasts until the cervix is completed dilated. Transition begins the period which moves labor from cervical change to ready for pushing.

Pushing: there is no longer a cervix to dilate. Contractions now work to expel the baby down and out.

Placenta: Labor is not complete until the placenta has been birthed. In some cultures time of birth is recorded when the placenta is born.

Coping and Comfort Measures

Labor Positions

Position is one the most effective birth tools one can have. There is an adage common in birth, "If you can't move the baby, move the mom." The knowledge and use of proper positioning can ease pain and position baby in such a ways as to speed the labor process. Correct positioning can encourage direct application of the baby's head to the cervix, and dilation will ensue.

 **Explore More:**

Research and practice the following positions. Find clear pictures to demonstrate each position.

© GoMidwife 2015

Create a hand-out to explain and depict each position and how it is used to advance and ease labor.

Lunge
Side-Lying
Hands and Knees
Standing
Squatting
Sitting on Toilet
Leaning

Movement

Movement accompanies and is similar to position. Swaying and walking are two way to integrate movement into labor, the the real significance the act of movement itself. It is easy to settle into one position and not move at all. Constant movement is not the goal, but rather a continual movement and change in position.

Breathing

There are many directed breathing techniques, but a normal, easy breathing pattern with the occasional deep breath in is best. A woman may need to be reminded to actually breath from time to time during labor. Encourage each mother to practice closing their eyes breathing in deeply then slowly let that breath out from tip to toe. This does not need to be done with every breath, but a few times in labor cleansing breathes can be refreshing to both mom and baby.

Massage

Massage can be used in various ways and methods. Firm massage can relieve tension in the shoulders or a cramp in the hip or thigh, whereas a light gentle touch can bring comfort.

Discuss various massage techniques in class. Each mother should determine which methods she prefers and inform her support team.

Hot and Cold

The use of both hot and cold in birth is an exceptional tool available in any environment. Hot compresses on the lower back works wonders for back labor or during the time when the baby's head is pressing against the lower pelvis near the end of active labor, warm showers can be relaxing in earlier stages of labor, and cold wash cloths with a hint of essential oil can be invigorating when labor is at its most intense.

Pressure

Simple counter pressure is one of the most efficient labor tools. Any one can provide counter pressure. Usually a mother prefers firm consistent pressure on her lower back during the active stage of labor. Sometimes pressure provided to her hips works best.

Hydrotherapy

Hydrotherapy is the use of water to bring comfort. This can be a simple shower or an actual birth pool in which to labor. Many women find relief in a birth pool when they cannot find relief any other way. Birth pools allow for the mother to be buoyant, easily incorporating movement into the labor naturally.

Aromatherapy

Smell is a powerful sense. Utilizing this in birth can bring great comfort, relief and even a burst of energy when needed. It can be used in the birth pool, on the cool cloths or in massage oil.

Environment

The birth environment must not be overlook as a birthing tool. If the environment is stiff, the it is understandable if the mom is too. Dim lighting, warm but not stifling temperature, familiar items and scents that stimulate or evoke comfort and safety should be considered.

Audience

The people who attend the birth will make a difference in how a woman births. The audience should never be discounted when it comes to birth preparation. Each mother should choose a limited audience and they should each support her birth philosophy. A person with great love for the mother, but concern and fear will unintentionally inhibit the birthing process.

Rhythm

Rhythm can be a comforting tool. A mother can completely relax between each contraction, but immediately assume a rhythm once she is aware the contraction is beginning. Often a mother cannot determine what her rhythm will be, but rather falls into it. Perhaps she moves to her knees places her hands on her hips and nods her head up and down, or she closes her eyes and sways back and forth, back and forth. Whatever her rhythm might be, encourage her to recognize and embrace it.

Rebozo

A rebozo is a piece of material used to "sift" the laboring belly back and forth or to lift the belly during a contraction. It is commonly employed to help encourage baby into a better position in the pelvis. This tool is more often used by a doula, but could be used by any birth support partner.

Muscles send messages to each other. Clenched fists, a tight mouth, a furrowed brow, all send signals to the birth-passage muscles, the very ones that need to be loosened. Opening up to relax these upper-body parts relaxes the lower ones.

William and Martha Sears

 **Enrichment:**

Create an informational on the physiology and stages of labor. This should be a visual guide prepared with your audience of women in mind. This can be, but is not limited to a hand-out. Always consider the best method of informational delivery. It is not always written text on paper. It should include: each stage of labor, what is occurring in the mother's body, what is the baby doing during each stage, length of each stage, length of contractions within each stage, the cardinal movements and overall estimated length of labor.

Create an informational on how position, movement and comfort measures help ease and advance the stages of labor. This can be, but is not limited to, a hand-out. Always consider the best method of information delivery. Consider all the learning styles. This is an area of teaching that can be very hands on and bring movement, engagement and enjoyment to your class. Think about how you could make an interactive activity.

Discussing the Birth Plan

By this point in the childbirth class, each mother should have considered and formed an idea of expectations, preferences and desires for how birth will go. Now is the time to articulate it in writing. Encourage each couple to take the time to write out what they desire for their birth. Here is what they should consider as they write:

- Who will attend the birth?
- When does she want to transition to hospital or call the midwife?

- Does she want to labor early on with her husband?
- Will she have a doula assisted birth?
- Which comfort measure will bring the most comfort?
- Where does she plan to give birth? If outside the home, when do they plan to transition over?
- Do you prefer a certain position to give birth?
- Will she labor in the tub? Give birth in the tub?
- If she is in the hospital will they use only a doppler?
- What foods does she prefer?
- What pain medicines will she accept?
- Will she decline an episiotomy? (This does still need to be stated and stated clearly.)
- Will she accept an epidural are ask that it not be offered.
- If a home birth is chosen what is the plan for transport in the event of an emergency?
- Who will watch other children?
- What are her feelings regarding a cesarean?
- What about delayed cord clamping and how does her practitioner define "delayed"?
- What are her thoughts on skin-to-skin?

 **Explore More:**

Look at the following example here: http://images.thebump.com/tools/pdfs/birth_plan.pdf and here: http://www.earthmamaangelbaby.com/free-birth-plan Considering your audience, create a birth plan form.

Contractions are an early sign of active labor — except when they aren't.
whattoexpect.com

Signs of Labor

Labor is a series of events increasing in progression. They can begin a week or even weeks before the actual arrival of the baby. Most women will show some sign of labor prior to the culmination of expectations. It is almost impossible not to become excited when these events begin to occur, but encourage each mother to take a deep breath and rest. There is, most likely. still a long way to go when these sign begin.

Here are some of the first signs labor is beginning to stir:

Loss of mucus plug: because this is one of the first signs it can, and often does, occur weeks before baby arrives. The mucus plug covers the opening of the cervix and protects from foreign substances contaminating the sterile environment and causing an infection. When the mucus plug falls out it means the cervix is beginning to thin and open. The mucus plug is slimy and sticky in nature, very much like the mucous from blowing your nose. It often is clear in color or tinged with pink.

Lightening: this is one of the first movements baby will make on his or her way out. There are seven movements considered to be the cardinal movements of birth, but lightening should be included. Lightening is also considered engagement. This means the baby's moves into the mom's pelvis. This action often takes place about two, even more, weeks before labor for first time moms.

Increased Braxton Hicks Contractions: prior to the onset of labor the uterus will usually be slightly more irritable. Contractions will come and go and there is often an increase in uterine activity as this muscle warms up for action. These contractions do not increase in intensity nor

© GoMidwife 2015

to they maintain a progressive pattern. If the mother's activity changes, they generally dissipate.

Increased Vaginal Discharge: preceding the onset of labor many women will experience an increase in vaginal discharge. It is not uncommon for this discharge to be mistaken for the mucus plug. This discharge should be mainly clear and not be accompanied by a foul odor or any kind of agitation.

Diarrhea: the happy birth hormones which begin to call the uterus into action also causes the bowels to stir. Early labor hormones can often have a cleansing effect.

Back Ache: when contractions wrap from the front of the abdomen around to the back, then the action is getting serious.

Rupture of Membranes: the bag of waters breaking is a true sign of impending labor. Most often the waters do not break until labor is well progressed, but it can happen in the very beginning and the action itself is the initiator of labor.

Increased Contractions: regular contractions that do not come and go, increase in intensity, and maintain a pattern with shorter intervals is true labor.

Remember 411: contractions that come every 4 minutes, lasting 1 minute for at least 1 hour

Warning Signs

Although designed to work and work perfectly, there are times when complications do arise. It is important for every mother to know her body well, be aware of these signs, what they indicate and how to respond.

Enrichment:

Research the following warning signs. Create a hand-out to explain each warning signs and what they indicate in pregnancy.

Bleeding Early Pregnancy

Bleeding Later Pregnancy

Headache

Decreased Fetal Movement

Contractions Prior to 37 Weeks

Abdominal Pain or Tenderness

Pain or Burning with Urination

Swelling

Severe Itching

Ruptured Membranes Prior to 37 Weeks

Blurred Vision

Severe Nausea and Vomiting

Fetal Proteins, Development and the Indication of Labor

Read the follow journal article:

Molecular mechanisms within fetal lungs initiate labor

June 22, 2015

UT Southwestern Medical Center

Researchers have identified two proteins in a fetus' lungs responsible for initiating the labor process, providing potential new targets for preventing preterm birth. They discovered that the proteins SRC-1 and SRC-2 activate genes inside the fetus' lungs near full term, leading to an inflammatory response in the mother's uterus that initiates labor. UT Southwestern researchers found that the proteins SRC-1 and SRC-2 activate genes inside the fetus' lungs near full term, resulting in an increased production of surfactant components, surfactant protein A (SP-A), and platelet-activating factor (PAF). Both SP-A and PAF are then secreted by the fetus' lungs into the amniotic fluid, leading to an inflammatory response in the mother's uterus that initiates labor.

Researchers at UT Southwestern Medical Center have identified two proteins in a fetus' lungs responsible for initiating the labor process, providing potential new targets for preventing preterm birth.

Previous studies have suggested that signals from the fetus initiate the birth process, but the precise molecular mechanisms that lead to labor remained unclear. UT Southwestern biochemists studying mouse models found that the two proteins − steroid receptor coactivators 1 and 2 (SRC-1 and SRC-2) -- control genes for pulmonary surfactant components that promote the initiation of labor. Surfactant is a substance released from the fetus' lungs just prior to birth that is essential for normal breathing outside the womb.

"Our study provides compelling evidence that the fetus regulates the timing of its birth, and that this control occurs after these two gene regulatory proteins − SRC-1 and SRC-2 − increase the production of surfactant components, surfactant protein A and platelet activating factor," said senior author Dr. Carole Mendelson, Professor of Biochemistry, and Obstetrics and Gynecology at UT Southwestern.

"By understanding the factors and pathways that initiate normal-term labor at 40 weeks, we can gain more insight into how to prevent preterm labor," said Dr. Mendelson, Director of the North Texas March of Dimes Birth Defects Center at UT Southwestern.

Each year about one in every nine infants in the United States is born preterm (before 37 weeks), according to the Centers for Disease Control and Prevention. Premature birth can cause brain hemorrhage and respiratory distress for babies, as well as long-term conditions such as cerebral palsy, chronic lung disease, and impaired vision.

The study, which appears in the Journal of Clinical Investigation, was supported by the National Institutes of Health and a Prematurity Research Initiative grant from the March of Dimes Foundation.

UT Southwestern researchers found that the proteins SRC-1 and SRC-2 activate genes inside the fetus' lungs near full term, resulting in an increased production of surfactant components, surfactant protein A (SP-A), and platelet-activating factor (PAF). Both SP-A and PAF are then secreted by the fetus' lungs into the amniotic fluid, leading to an inflammatory response in the mother's uterus that initiates labor.

The current study showed that a deficiency of both SRC-1 and SRC-2 inside the fetus' lungs drastically decreased the production of SP-A and PAF, causing a one- to two-day labor delay in mouse models, comparable to a three- to four-week labor delay in women.

Researchers further found that injecting either SP-A or PAF into the amniotic fluid of the deficient mice allowed the mothers to deliver on time. Together, the findings further define the underlying molecular mechanisms by which fetuses control the timing of birth.

Future research will include defining how fetal signals are transmitted to the mother's uterus, and relating these findings to the causes of preterm labor.

The study was conducted with current and former UT Southwestern researchers, including first author Dr. Lu Gao; Dr. Elizabeth Rabbitt; Dr. Jennifer Condon; Dr. Nora Renthal; Dr. John Johnston; Dr. Matthew Mitsche; and researchers from the Institut de Génétique et de Biologie Moléculaire et Cellulaire, France, and Baylor College of Medicine in Houston.[17]

This article clearly shows the development of the proteins within the fetal lungs that initiate labor. If labor is not initiated then that should clearly show the baby is not ready. There is so much concern with a baby growing too big or being post dates and this argument is usually the one used to convince mothers and fathers that an induction is necessary. This article, however, indicates that the induction process excludes one very very important member of the labor process: the baby. Encourage your mothers to understand what they are agreeing to, and why, when they agree to an induction. An induction, is still a choice.

Induction

Induction was discussed early as an intervention, but it is imperative to also discuss it here as a right. It is the right of the mother to choose not to be induced. It is essential tor women to understand the difference between when an induction is medically necessary and when they are faced with an induction brought about by fear of litigation. All too often hospitals will induce because their risk of liability risk is lower when a woman is induced early versus when the pregnancy is left to progress to what is called "post dates". Litigation, or the potential of it, is never a reason to induce. It is a woman's right to say no. The rates of induction have risen dramatically in the last few decades. Often it is easy to convince a woman an induction is a good thing. By 40 weeks everyone is just ready to be done. They are hot, uncomfortable and sleep is elusive, but these last few weeks and days are necessary and we should encourage women to appreciate them

Birth Rights

One of the important concepts in pregnancy and birth, and one that is often not discussed, is the issue of rights. We often think about rights being something that people are denied in other places, far away from the comforts of our homes and lives. The truth is that women in the developed world also have basic human rights that are often infringed upon. For our purposes, the discussion of rights really focuses on the ability of the woman to choose. The choices she is able to make include when and how her baby will be born. There is no reason to allow these choices to be made by someone else. A doctor might have more training and certainly might have an opinion about when and why a certain procedure should be done, but in the end it is the mother's choice. There are obviously medical emergencies when this can change, but the need for induction or being strapped to a monitor is the choice of the mother. This includes medicines she might choose to take or refuse, procedures such as episiotomies that may be unwanted, or any number of other things. The key is to inform the women you serve on what their rights are and help them to learn methods to exercise them. The use of a doula can be helpful in this regard as they can advocate for the mother.[18]

Notes

Excellence is never an accident. It is always the result of high intention, sincere effort, and intelligent execution; it represents the wise choice of many alternatives - choice, not chance, determines your destiny.
Aristotle

Self Education and Responsibility

Informed Choice

Within western medicine the societal implications are that a doctor has implicit authority within the setting of their office or hospital. For this reason alone, there has been an increase in home birth as a choice for women to give birth. Some would argue there is an increased risk in home birth, but for many women that minimal risk far outweighs the impotency they would have under the totalitarian medical system. To maintain authority over decisions in birth and their body, they accept the implied risk and they give birth outside of the system. This is informed choice. They have been informed of their options. They understand and have evaluated each alternative, the risks and benefits and have made an informed decision. Blind allegiance to a broken system should no longer be tolerated. Women must be informed of their options and, with an educated understanding, make a choice of what is right for their family. Once they choose, whether or not we agree with that choice, if they have been informed, we must support each woman's decision as educators and birth workers.

Informed Consent

After the overarching decision has been made of where to give birth, it should be understood there will still be many choices to be made. No one can administer any medical procedure without expressed consent. As the educator, you can only bring a mother so far; it will ultimately be up to her to determine where the boundary lies for her. You must educate her on what the possibilities are, and she must determine how she wants to walk her birth options through. Encourage her to understand if she has written a birth plan, and there is no medical necessity to her or the baby for a particular procedure or medication then regardless of facility type she has a choice to say, "No". It is your job as an educator to inform her of her choice. It is hers to consent.

Packing Your Birth Bag

When packing for the hospital it is necessary to consider all the items you or your partner might need as there won't be the flexibility to grab forgotten items, especially if the hospital is far away from your home. A personal opinion is to err on the side of items you won't necessarily use but might won't versus under packing and wishing you had brought that item with you. The good news is a hospital stay is generally only a day or two and you will be quickly back amongst your things before you know it. Here are some items to consider:

For the Hospital

Your birth plan
ID and insurance cards
Change for the vending machines

Toiletries and Personal Items

Toothbrush and toothpaste
Shampoo
Body wash
Deodorant
Witch Hazel pads or PP bottom spritz
Glasses and/or contact case and solution
Make- up (should she choose to primp prior to departure)
Robe
Tube top or bikini for a water labor
Phone/camera
Nursing bra

Open front shirt

Comfortable clothes to return home in

Nightgown

Cozy socks

Cell phone charger

For Labor

Chapstick

Hair bands

Easy to digest snacks (check with the facility on whether or not she can consume these foods)

Honey sticks for a boost of energy

A misting fan

Music playlist and speakers

Massage oils

For Baby

Car seat

Clothes (soft and comfortable)

Diapers

Hat

Preparing Your Home

Should a woman choose a home birth she will need to prepare her home to receive both the birth team and her newborn. There are many matters to address and she will need to begin this long before the expected arrival date. Here are some ideas for her to consider:

Children in Home Birth

Many families who choose to birth at home desire for siblings to be present or freely able to come and go within the birth room. This is your home and your birth, and children are welcome to participate as you see fit. We will welcome your children into your birth with you as long as they do not become a concern for the laboring mother. We believe it is vital that if the next generation is to embrace birth as normal, they must see it as it is. We have never experienced children afraid in the birthing process. Most intuitively understand, but during prenatal visits and as you walk with them throughout your pregnancy, it is important to talk to your children about what to expect.

As your birth team, we do ask for you to have a designated care giver in the home or available to come at notice to care for the children. Although most children thrive in the birthing environment, it can be taxing for the mother and prevent her full surrender. There are also times, rare but possible, when a transfer is necessary. Both the father and birth team will need to accompany the mother, and young children cannot be left unattended. Please choose someone who will be able to care for your children during your birth.

Birth Tubs

If you have chosen to give birth in water it is important to ensure proper care of the tub prior to birth and the understanding of how to set it up. We encourage you to do a test run prior to birth. Not all hot water heaters are created equal and it is important to know you have the proper adapter and length of hose prior to the moment of necessity.

We strongly encourage you not to fill the tub and leave the water standing. Standing water should be drained every 48 hours. We would encourage you to drain the tub after every use. If, when doing the test run, you want to soak or float awhile then that is fine, but be sure to empty and clean the tub prior to the day of birth. Some mamas here in Hawaii have catchment water. Catchment water is considered safe, but we do encourage 1/2-1 cup of bleach be added to the water to ensure bacteria does not grow. You will need to stir the pool and and allow it to settle for at least 2 hours. If you do not want to use bleach then you can use 2 cups of food grade hydrogen peroxide.

If this is your first baby then you should consider having your spouse fill the tub when contractions are approximately 4 minutes apart for about an hour and lasting 1 full minute. Tubs are considered the "aquaderal" of home birth and provide immense relief from labor pains and are especially helpful for first time moms.

What to do in Early Labor

Notify your midwife and doula when you think you are in labor.

REST! It cannot be stated enough that rest is essential to successful labors and home births, especially for first time moms. When labor starts and you can still walk and talk through a contraction then rest is the most important task of the laboring mother.
Eat a good protein in early labor. You will not feel like eating as labor progresses and it is important to be nourished for the journey ahead.

Stay hydrated!

Make your bed. We ask for you to make your bed and then place a shower curtain or plastic over those sheets and make your bed with an old set of sheets on top of the plastic. If you should choose to have the baby in bed we can easily strip your sheets and tuck you and the baby into the fresh ones on the bottom during the postpartum period.

Breathe.

Enjoy! Often mother's miss this aspect. Birth is to be celebrated. Sure, it is hard, hard work, but as women most of us will only ever experience birth a handful of times, more if we are fortunate. Enjoy the moment. Remember, your baby is working hard, too. Focus on your baby and not the pain.

Materials for You to Gather:

When gathering the needed materials for a home birth, consider purchasing two laundry baskets. Gather all the supplies and put them in one place near where she plans to deliver. This way all materials can be gathered quickly and moved to another location if she choose to birth in another room.

Birth Kit or Water Birth Kit Ordered

For Birth:

2 large trash bags
Laundry basket
Bowl for placenta
Mattress protection
Floor protection- plastic drop cloth
Hydrogen peroxide
Box of disposable gloves
Large cookie sheet
5-10 towels
5-10 wash cloths

For Comfort

Popsicles made with fresh fruit and coconut water. These are refreshing, easily digestible and can add nourishment and electrolytes while in labor.
Drinking Straws
Candles
Music
Robe
Healthy snacks
Fan

For Postpartum

Gallon freezer bag
Large diapers (Depends)
1 Package panty liners
Robe

For Baby

Cloth diapers
Wipes
Car seat *secured in the car*
5 baby blankets

Unmet Expectations

We prepare women for the best and most ideal outcomes but, sadly, not every birth will rise to their greatest expectations. This can be heartbreaking, especially when she has planned a beautiful, natural birth, laboring at her own pace with music playing soothingly in the background and she arrives at the exact opposite and without much, if any, preparation. It is hard for identity not to be woven into these lost hopes and for perceived failure not to be assumed. Discussing this in class is very important for her to be able to articulate how she might plan to respond or cope should drastic changes in the plan take place. Most women will understand if circumstances change, but it's often the abruptness of that change which can be devastating. It would not be balanced if, as her educator, you did not broach the subject. Also, should you know of a student who does have a birth completely different from her plan then follow up with her. Continue to be her resource and expand her network. She will need even more postpartum support. Spend time listening to her and validating her unmet expectations. Be mindful of your responses and encourage her to embrace the beauty in her arms. Birth is birth and no matter how we envision it, the end is more important than the means. It is okay to grieve the loss of dreams, but it in no way lessons that woman as a person or a mother.

Grief and loss

It is paramount to recognize, unless you have specialized training, that you are not a grief counselor; you are an educator. As her educator you may be her only resource. Be prepared to facilitate her connection to proper professionals who can continue her care. Understanding the cycle of grief will more readily prepare you for assessing a woman's needs and help you determine how best to provide proper materials and recommendations.

Grief is normal. It is to be legitimized. Our culture is afraid to embrace those who grieve. Perhaps we are afraid because we consider loss contagious. We are ourselves casualties when we do not confront truth in regards to the human grieving process. We do not know how to best comfort those who have experienced, or are experiencing, grief because like birth, grief is hidden away. If we do not comfort those who are mourning now, we ourselves will not know how to be comforted. As the educator of one who has or is experiencing lost, do not be afraid to be real. Guard your words and be slow to speak, but do not abandon her.

 **Explore More:**

Read about grief and the process of recovery.

http://www.goodtherapy.org/therapy-for-grief.html
http://www.helpguide.org/articles/grief-loss/coping-with-grief-and-loss.htm
http://www.griefhealing.com/column-understanding-the-grief-process.htm
http://www.recover-from-grief.com/stages-of-grief.html
http://psychcentral.com/lib/the-5-stages-of-loss-and-grief/
http://psychcentral.com/blog/archives/2011/11/20/8-tips-to-help-console-a-grieving-friend/
http://www.healthy-holistic-living.com/words-to-comfort-someone-grieving.html
http://www.helpguide.org/articles/grief-loss/supporting-a-grieving-person.htm

Of all the rights of women, the greatest is to be a mother.
Lin Yutang

Postpartum and Newborn

It would be an understatement to say postpartum is an overlooked period in the childbearing years. It is in fact, all but dismissed as part of the process. According to words, not true, but according to the actions of society childbearing ends in childbirth, and women should be self-sufficient there after. Postpartum is however, one of the most significant times and for many reasons. One of those reasons is this postpartum period will effect her next pregnancy. How she recovers physically, emotionally and spiritually will impact future pregnancies.

In the developing world, the postpartum period is when mothers and babies die from preventable infections because there is little or no postpartum follow-up. In more advanced societies women will suffer from increased postpartum depression and poor recovery because they are hormonally imbalanced, emotionally fragile and physically spent with little or no support. This period can make great impacts on the lives of women and door is wide open for improvements in care to be made.

As an educator, it is your job to make women aware of these possibilities. Just as when women go into labor unprepared they fall into poor outcomes, so too poor outcomes occur in postpartum because women have little insight into this unknown. Encourage each mother to set up a postpartum support team with supporters equal to her birth support team. If you, as an educator, work with a sector of women without this resource availability, consider forming a community postpartum support team.

 **Explore More:**

Consider what a community postpartum support team might look like. Who could be involved? How would it be set up? What would it incorporate?

Warning Signs for Mother

Bleeding: Bleeding is part of the postpartum process. Heavy bleeding is not. If a woman fills a heavy menstrual pad in and hour or less, bleeding is considered profuse. This is postpartum hemorrhage levels of bleeding. Not common, but postpartum hemorrhages can occur up to 2 weeks after birth.

Extreme Sadness: Efficient emotional and physical support in the postpartum period will inhibit postpartum depression, but it can still occur regardless of the support a woman receives. Baby blues do occur with hormonal shifts, but continued, persistent severe sadness is not normal.

Fever: A slight increase in temperature can be benign, but it can also be a sign of an infection. If a fever is accompanied by chills, tenderness in the breast, abdomen or perineum, or if there is foul smelling discharge present she should see help.

Pain in Perineum: There will be typical soreness for a few days preceding birth, but severe pain or tenderness is not normal.

Odor: Foul smelling discharge is a sign of infection.

Tenderness in Breast: A tender, red or sore area of the breast can be a clogged duct and relieved by warm compresses or expressing. Pain that is not relieved or accompanied by red streaking or flu-like symptoms is likely a breast infection and will need an antibiotic of some kind.

Pain in Leg: Hormones increase a woman's clotting factors so she will not bleed excessively after birth. These increased factors can lead to a clot in the leg. If pain, redness, swelling, tenderness occurs she should seek help.

Headache or Blurred Vision:

Pre-eclampsia is not limited to pregnancy, but can also occur in the postpartum period.

Seek help immediately if...

There is shortness of breath

There is heavy bleeding

She feels light-headedness and it is accompanied by any of the following: rapid heartbeat or palpitations, rapid or shallow breathing, clammy skin, restlessness or confusion.

Physiology of Postpartum

Just as it took time for the physiology, or function, of pregnancy to take place so it will be for postpartum recovery. Bodies are designed to functional appropriately, but it is always nice to help them along with proper nutrition and rest. Here is what to expect during the postpartum years (yes, years).

It takes approximately 6 weeks to recover from the initial effects of pregnancy and birth on a mother's body, but it can take 18-24 months until the body has completely recovered and is prepared to birth again. During the child bearing years a mother is either ante (before), intra (inside of) or post (after) partum (childbirth) at all times. The time a mother generally nurses her child is about this amount of time as well, and nursing can prevent pregnancy so again, the design is perfect. Once the child has weaned, it is time to do it again and a woman moves from post to antepartum.

Early Weeks Postpartum

Milk- Milk should come in around 2-4 days postpartum.

Weight- Following birth a mother can expect to lose 8-12 pounds of baby, placenta, fluid and uterine weight. Within the first few weeks she should expect to lose another 3-4 pounds.

Involution of the Uterus- Immediately following birth the uterus can be felt at the umbilicus, or the belly button. It should recede 1-2 finger-breaths each day after birth disappearing behind the pubic bone around 2-4 weeks postpartum.

Cramps- The uterus will continue to cramp, or contract in the days and weeks postpartum. This cramping is what shrinks the uterus back down to pre-pregnancy size. A warm rice sock or heating pad can help. Red Raspberry Leaf tea is good too as well as After Ease Herbs.

Belly Muscles- The abdominal muscles separated to allow room for baby to grow, and will now need to come back together. Belly-binding can help these muscles come back together, a process that will likely occur naturally even without belly-binding.

Exercise- This really is on an individual bases. If a mother exercised throughout her pregnancy then she can resume similar exercise soon after birth. Two weeks postpartum is probably a good rest period and even a full month if possible. Exercise is a good thing, and it can help speed recovery if a mother is already used to it.

Sex- As with exercise, this is an individual decision. Bleeding should have stopped prior to intercourse, but after that consideration, which is usually 2-4 weeks postpartum then it really is up to how she feels emotionally and physically. Like exercise, sex is a good thing and a return to routine can increase postpartum recovery.

Herbal Bottom Spray

Ingredients

- 4 oz bottle with sprayer
- 2 oz witch hazel
- 1 tsp vegetable glycerin
- 2 TBS aloe vera gel
- purified or distilled water
- 10 drops peppermint, lavender, rosemary and frankincense essential oil

- 5 Drops Coconut Oil

Instructions

1. Fill a sterilized 4 oz bottle halfway with witch hazel.
2. Add glycerin, aloe, essential oils and coconut oil.
3. Fill the remaining space in the bottle with water.
4. Shake well and place sprayer lid on bottle.
5. To use: shake well and spray onto affected areas as needed.

Expectations

Expectations in postpartum need to be full of grace. The length of time she will bleed, when she will feel like having sex again, even when she will go a whole day without feeling weepy. Like birth, postpartum is varied. Some sail through and love every moment, others labor. Encourage her to set realistic expectations for herself and her home. Nothing needs to be rushed; this is a precious time of bonding and learning and growing together as a new, larger family. Help family members to understand the need to make memories and enjoy every moment as it will soon be over. There will be many opportunities to get back to a normal routine, so set this time apart to all learn with one another the new ways this family will work. Rushing back into anything or hoping to get past a certain milestone only puts pressure and unwanted expectations on everyone. Do not allow your new mother to be set up for disappointment!

The Social Network

Postpartum can be a precarious time. Often mothers feel, or are, left alone. The climax, as society sees it, is over. The birth has occurred. Mom is now hormonal, leaky, exhausted and overwhelmed. A mother well prepared for labor and birth may not at all be prepared for the postpartum period. She needs to understand this PRIOR TO the postpartum period and plan her postpartum team just as diligently as she planned her birth team. A quick and personal story: I have literally served women whom I brought two assistants to her birth and then needed seven assistants during her postpartum period. She rocked her birth and labored during her postpartum. Labor occurs many times following conception… not just during the 24 hours surrounding birth. It is imperative a woman has a social network who can bring her meals, drive her around in the sunshine with the windows down, watch other children or simply just sit with her and listen. This time is often overlooked by both the mom and her team and it is one that will make a significant difference in her mental, emotional and physical recovery.

Safe Sex

When resuming sexual activity after birth it is necessary to consider a few things:

- Discuss the return to sexual activity with her partner.
- Sex after birth is normal. It can and should resume.
- Wait until bleeding stops.
- Be gentle and slow the first time.
- Use a lot of lubrication, preferably a natural lube such as coconut oil.
- There may be some pain with initial intercourse, but it should not continue. If it does it could be due to the way a tear has healed. Seek attention if it persists.
- Often women will have trouble with either desire or function from about 3 months (when most return to activity) until about 6 months.
- A woman can get pregnant as soon as she ovulates, which can be as early as 6-8 weeks postpartum.
- The birth of a baby changes the relationship between a mother and father, but never needs to reduce the relationship.

Family Planning

Child spacing is essential for the healthy family and healthy mother. There are increased risk to both mother and baby with pregnancies too close together. If this was not a planned pregnancy, it will be important to plan the next one. If this was planned, then now is the time to consider long term family plans and goals. It is also necessary to now consider, if it hasn't been already, what her family planning philosophy is. This information will be key in deciding how to plan accordingly. In the developed world there are several apps that can be used for family planning, one of the best being *Fertility Friend*. In the developing world fertility beads might be used instead.

 **Explore More:**

Research the following methods of fertility, all of which are stated safe during breastfeeding. Determine which method is best and why based on your on philosophy of birth and family planning. Create a form that explains each method, and the benefits and concerns of each.

- NFP
- Cervical Mucus Method
- Condoms
- Diaphragm or cervical cap
- The mini-pill
- Injection
- Implant
- Intrauterine system (IUS)
- Intrauterine device (IUD)

Feminine Hygiene

Feminine hygiene in first world nations need very little discussion. One point would be the use of a peri-bottle with an herbal rinse in the early postpartum days. This will clean the area after urination, but not cause additional soreness after birth from wiping. In developing nations however, this will need to be a key point of discussion. Vaginal infections can quickly ensue in postpartum if a mother does not maintain proper hygiene. Where water is scarce or sanitary pads not available this risk is heightened. It is in no way to be indicated they choose poorer hygiene, in many cases they have never been educated in the necessity for or how to properly clean themselves. Education on basic hygiene is a key to reducing postpartum infections.

Explore More:

If you plan to work in a developing nation create a teaching to demonstrate and explain basic feminine hygiene. Remember, if you are in another nation language and culture will be a barrier. Consider how to express this teaching without words.

Notes:
_____
_____
_____
_____
_____
_____
_____
_____
_____
_____
_____
_____
_____
_____
_____
_____
_____
_____

For this child, I prayed.

Samuel 1:27

Warning Signs for Newborn

Babies are designed to thrive, but there are times when complications arise. When they do, it is important for new mothers to know what is normal, what is not, and when to seek help.

Action	What is Normal	What is Not
Crying	Babies cry. It is how they speak to us, but they should calm when comforted, held and tended to.	Crying that cannot be comforted. High pitched, squealing cry. Sudden cry with a marked difference.
Eating	In the first few hours and days of life, babies are just beginning to establish patterns. They may eat following birth then take a long recovery period where they mainly choose to sleep. After the first few days a baby should be eating about every 3 hours.	A poor appetite. Not waking to eat. Not eating when awakened. Poor suction. Vomiting after nursing.
Sleeping	Babies sleep about 18 hours per day.	Not waking to nurse after the first 24-48 hours. Lethargy. Poor energy when awake; listless. Not alert when awakened.
Belly Button	Crusty, dry, with a little dark dried blood noticeable.	Wet, smelly, oozing, streaked with red, warm.
Pees and Poops	4+ wet diapers per day. At least one movement for the first few days, increasing to 6+ per day. Babies will establish their own pooping routine and do not have to have a movement every day after the first few weeks of life, but should have wet diapers daily. Prior to milk coming in (the first 4 days of life) it is common to see pees tinted with orange, pink or red. This is due to concentrated urine crystals.	Less than 4 wet diapers per day. No pooping in the first 24 hours of life.
Temperature	A baby cannot regulate their temperature. Skin-to-skin is best for maintaining a normal temperature. 96.8 to 98.6° F	If environment is not the cause of either a warm or cool baby, then seek attention. A baby with a low or high temperature could have an infection.
Breathing	A baby's breathing is irregular. They may take several rapid breaths then appear to not breath again for a couple -5 seconds. <60 per minute	Grunting or wheezing sounds. Retractions, or when the skin pulls in around the ribs or in the middle of the chest. Rapid breathing >60 per minute

© GoMidwife 2015

Normal Newborn Development

Head- A molded head is very common, especially in first time moms. This shape should move nicely to a beautiful round head within the first 72 hours of life.

Skin- Healthy babies have a myriad of skin rashes in the first weeks of life and most are perfectly normal. Jaundice, the yellowing of the skin, is also normal in breastfed babies. The yellow color should not look orange or extend below the belly button or to the extremities.

Weight- A healthy baby will lose about 10% in the first week of life, but they should weigh what they did at birth, or more, by the second week of life. After they, a 1-2 pounds gain should occur by the end of the first month.

Movement- The baby's first movements will be rooting for the breasts. They will likely open and close there fingers around your breasts and fingers. He or she will also stretch from time to time, but for the first 5-6 months baby will remain mostly content to be curled as they were in the womb.

Growth Spurts- It is common for a baby to have a growth spurt around 7-10 days, 2-3 weeks, 4-6 weeks, 3 months, 4 months, 6 months and 9 months. They will often increase their nursing as well as sleeping rhythms during this time.[19]

Breasts- Newborn girls can commonly have milky discharge from her nipples. This is due to her mother's hormones. It is known as "Witches Milk."

Genitals- It is not uncommon for the genitals to be swollen in the first hours, even days after birth. Also, not uncommon is for little girls to have discharge and blood from her vagina. This is due to the mother's hormones she received in the womb.

 **Explore More:**

Research and create a handout about normal newborn development, behavior and what a mother should expect in the first days, weeks and months of life. This should be a hand-out they can take home as a reference. Be sure it is informative, clear and engaging.

Welcome to Parenthood

What do they need to know as new parents? First of all, it is important to know it will be okay if it is not perfect. Just as it has been discussed the need to approach child birth with the understanding all three components of our being are engaged in the process: the physical, emotional and spiritual so it is with child rearing. As with the birth there were moments of struggle and discouragement as were there moments of great triumph; so it will be with a newborn and a child. Encourage her to trust her instincts, they will be there, to believe in herself and to surrender to the process.

Here are some thoughts to consider:

- Will she circumcise?
- Will she immunize?
- Who will be the baby's doctor?
- Will she breastfeed exclusively?
- Does she have a support team for meals?

Then remember:

- Sleep when she can. If baby sleeps, she sleeps no matter what needs to be done.
- Hold baby gently, supporting the head and neck and never shake.
- If frustration creeps in seek a friend or helper.
- It is okay of the house is messy or laundry is not washed.
- Breastfeeding is hard work, but it is the healthiest for her newborn and herself and it is economical.
- Remember when soothing a fussy baby where she or he lived for nine months: in a dark warm environment with gentle swaying movement.
- Enjoy the moments.

Consider:

As an educator, facilitate a mother's group or a parent's group. An informal group where mothers and/or fathers can come and relate to their peers, enjoy a snack have a discussion surrounding an appropriate subject for parenting and perhaps receive a 5 or 10 minute teaching from an older mother, yourself or another educator. This would allow you to continue to be a part of their lives and guide them through the early years. In the developing world, the maintaining of relationship is vital to improved outcomes and for future pregnancies. This is an opportunity to continue to walk alongside families and serve them, and it provides them with a much needed outlet for discussion and support.

Notes

Breastfeeding is a natural "safety net" against the worst effects of poverty. If the child survives the first month of life (the most dangerous period of childhood) then for the next four months or so, exclusive breastfeeding goes a long way toward canceling out the health difference between being born into poverty and being born into affluence . . . It is almost as if breastfeeding takes the infant out of poverty for those first few months in order to give the child a fairer start in life and compensate for the injustice of the world into which it was born.

James P. Grant, former Executive Director, UNICEF

Breastfeeding

Breastfeeding is a fundamental foundation for healthy babies and healthy mothers. As with natural birth, breastfeeding has not been demonstrated as normal throughout the last few generation and women need to be educated on the fact it is the unprecedented answer to initiating healthy newborns. Like natural childbirth, breastfeeding is not embraced without controversy, but the more women know they more they will recognize its incomparable value.

 **Explore More:**

Using www.kellymom.com as at least one resource, create a hand-out explaining each of the following:

Why is breastfeeding important?
When does milk come in?
What is colostrum?
How will she know baby is getting enough milk?
What are the benefits to mother?
What are the benefits to baby?

What should she expect in the first few weeks of breastfeeding?

Why is breastfeeding even more important in developing world situations or lower economic settings?

Education is not preparation for life; education is life itself.
John Dewey

Community Classes

Classes should be offered on a consistent basis, especially in developing communities. It is quite possible, for the first few months, to only have one or two moms who take advantage of the classes you offer, but word of mouth is the best form of advertisement. As each mother tells her story of how education effected her birth outcome then classes will fill.

Example Schedule One	Example Schedule Two
Class Time: 2 hours Time: 7pm-9pm	Class Time: 2 hours 15 minutes Time: 6:45pm-9pm
Teaching Time Topic One: 15 min. **Discussion Time:** 10 min. **Hands-On Activity:** 15 min. **Teaching Time Topic Two:** 15 min. **Discussion Time:** 10 min. **Hands-On Activity:** 15 min. **Teaching Time Topic** Three: 15 min. **Discussion Time:** 10 min. **Hands-On Activity:** 15 min.	**Teaching Time Topic One:** 15 min. **Discussion Time:** 10 min. **Hands-On Activity:** 15 min. **Movie:** 1 hour 15 min. **Discussion:** 20 min.

Centering Pregnancy:

Consider offering centering pregnancy classes as an option to the educational approach. Centering pregnancy's objective is three-fold: health assessment, education, and support. Invite a midwife to join your educational group and offer health assessments as part of your objectives to education. Centering pregnancy also works in it offers a built-in peer-support system where women get to know other women in their community. This widens their support resources as well as allows them to network with and build peer relationship that may last a lifetime.

Individual Approach

Offering individual classes to one or two couples at a time is also a considerable way to approach education. Keeping the classes small and intimate allows each mother to have one-on-one attention with the ability to connect deeply with you as the educator. This approach will likely see you as part of her support team instead of a peer.

Intelligence is the ability to adapt.

Stephen Hawking

Adapting to the Needs of the Audience

Enough cannot be said about adapting to the needs of your audience. Consider the holes within the community you choose to teach. Does this area have high cesarean rates? Then offer a class specific to meet the needs of VBAC mothers. Does your community have women susceptible to malnutrition be it the inner city of Detroit or the rural wastelands of the Congo? Offer a course specifically focused on nutrition. Are you in the suburbs of Minneapolis where there is a large population of Somali women? Offer a course on birth after female circumcision. Recognize the needs of the people and strive to educate them in their need.

The single biggest problem in communication is the illusion that it has taken place.
George Bernard Shaw

Communication Skills

Language

Depending on the community you serve, language may be a barrier. This does not mean childbirth education is not an option, it just mean you will need to work to improve overall communication by learning what is not clear and overcoming the gap in understanding. When presenting to an audience where language is an impediment remember to speak slowly and keep it simple. Practice observing the class to know where they may not fully comprehend what you are saying. If necessary, ask if they need clarification and repeat what was said. Communication is about more than just the delivery facts and information, it is about successfully conveying ideas, understandings and concepts to another. When language s a barrier to communication watch for their non-verbal cues and be aware of their emotions regarding the matter. If language is a barrier, then most likely you are speaking to an audience of a different culture. Culture will make an appreciable difference.

Culture, Religion and Ethnicity

Techniques in communication are not universal. This means they do not necessarily translate as successful. One style in one culture may not work at all in another. For example, in many western communities education is lecture based, but in oral cultures it would not be out of the question to use a puppet show to teach valuable concepts. In a western society puppets would be considered for children, but in oral cultures they are accepted.

Learn about the culture you will be teaching. Even if they do understand English, there may be hand gestures, particular words or pictures which need to be eliminated from this particular class. Knowing the culture you plan to teach and formulating your material to them will give them the understanding you value them and their culture and they may be more willing to listen and receive what you have to share.

Effective Methods

When communicating with others it is fundamental to recognize the following keys in order of importance in communication technique:

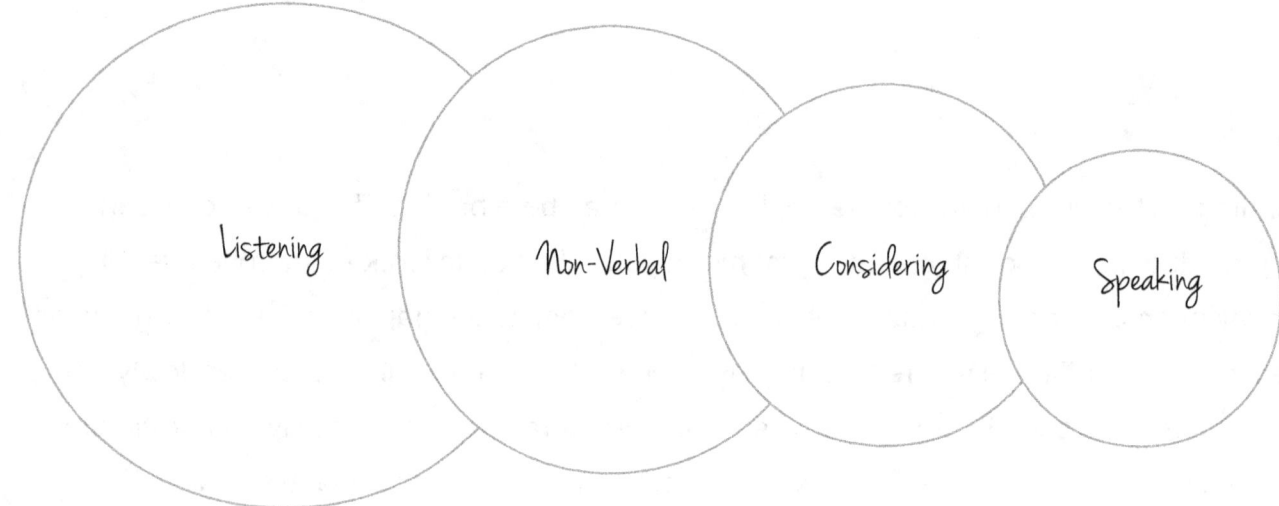

Listening-

Listen to those you will teach before you teach them. Actively hear what they have to say. This will give you keys to what their perspective is, what their needs are, how they learn and what they desire to know. At the beginning of each class take time to get to know them. Ask them to tell about themselves then ask pointed questions to learn more about them and how you can best teach them. As the class proceeds continue to listen and continue to ask questions. Listening is probably the number one key to good and effective communication.

Non- Verbal-

Non-verbal gestures, facial expressions and body language comes next in the order of importance when communicating on either an individual or group level. If you are speaking of something in a positive manner, but your face or body registers a negative response the audience will listen more carefully to what is not said with words, but rather what they see being said with non-verbal expression.

I like to listen. I have learned a great deal from listening carefully. Most people never listen.

Ernest Hemingway

Considering-

Consider what they already know and consider what they need to know. If you have listened them, then you will recognize how best to formulate your teaching to better accommodate them where they are currently. There is no need to go in-depth over VBAC and how to emotionally and physically prepare for birth after a cesarean if every mother you are teaching is a first time mom.

Speaking-

Speaking should be the least common way we communicate. Use words when necessary, but a whole lot of words are quickly forgotten. If you can evoke an emotion from them, then it will be worth more than a 1,000 words. Keep in mind more people learn from either doing or seeing, more so than hearing. Do not plan classes around beautifully orated soliloquies. Bring them into and make then a part of the education process and they will own it for themselves.

Notes_____
_____
_____
_____
_____
_____
_____
_____
_____

Teaching Strategies

strat·e·gy a plan of action

The strategy of how to approach teaching will vary from group to group. It will not be enough to formulate one given strategy, rather there will be need to reassess and re-strategize after each class session and at the beginning of each new group. In order to be effective in teaching you must reach your audience and actively cultivate their engagement. Even the slightest change in approach can make all the difference in whether or not the audience feels freedom to engage.

Here is an example of how a simple (and accidental) change modified the whole dynamic of a classroom and revolutionized audience participation.

I had been teaching childbirth educator classes for many years with success. My general approach for setting up class was in a formal classroom setting, as that was the facility available to me at the time. The students sat at long study tables with 2-3 chairs at a table and all facing the front of the room. I sat on a stool at the front of the class.

I had formulated an outline filled with hands-on activities, discussions, movies and lecture. To date this approach had been met with active participation, enjoyment, deep discussions and satisfied learners. However, all this changed with the arrival of one particular group. This group were all relatively young learners, in their late teens and early twenties. I went about the class as usually assuming all would go as designed and planned. However, I could not convince them to engage. Class after class I was a circus monkey on a tightrope before them. Each class I tried this actively and that, this new discussion, and that. Nothing. They stared blankly at me and nodded. No one spoke unless I directly asked an individual a question. I was devastated, and could not figure out the key to truly get these girls to open up. I would like to say I formulated a great planned and overcame the odds, but in truth I decided I would just ride it out until the program was over and mark this down as a failure. I just did not know what to do.

As it would happen the next week my usual classroom at the university was taken and I had no other options. So, I invited the girls to my home to hold class there. When they arrived I prepared some teaching and they all sat around my living room, some on couches, some in chairs and some on the floor. My lesson prepared in the usual way, I set out to encounter the same non-reaction as before and encouraged myself that there were only three sessions left. Only this time....I could not stop the girls from talking. They openly engaged, asked pointed questions, discussed amongst themselves, revisited with me all the bottled up questions from the 4 weeks before that they never asked. It was one of the best classes I have ever held. What was the difference? **Only the setting**.

This story depicts how powerful the simple details are to either make or break the atmosphere of learning, and in the end, the atmosphere of learning is equal too if not more compelling than any other single event within the classroom setting. If the atmosphere is not conducive to learning then they will not learn, and if they do not learn then birth outcomes will not change.

Strategies, or our plans of action, must always be transforming, organic, able to adapt and target how this particular group will learn. Teaching cannot be vanilla. It absolutely cannot be redundant, it must be molded and reconstructed, transformed and inspired The information may be the same, but the delivery must always be fresh and alive.

Notes _____
_____
_____
_____
_____
_____
_____
_____
_____
_____
_____
_____
_____

Simplicity is the ultimate sophistication.

Leonardo da Vinci

How to Set Up a Classroom

Before you determine how to set up your classroom, you will need to know where you will teach. Will you offer classes from your home and will the setting me small, intimate and non-formal or perhaps will you teach your classes at the local library, church or school? Non-formal, small, intimate settings seem to go the farthest in pulling down defenses and barriers, and opening one up to listen, participate and learn. It may be thought, in a childbirth education setting there should be little or no defenses, but childbirth is a highly emotional topic and often unintentional guards are present. Your goal in arranging the classroom set up is for the mother to be able to feel comfortable and relax. Within the intimate setting, keep it new with changes weekly to the surroundings. Although you want to evoke comfort, you do not want to settle into the habitual and the routine. Simply changing up the seating arrangement can become a teaching concept for childbirth: last week was comfortable and it worked, but this week let's try a new set-up this may prove to be even better than the last "position".

Classroom Interaction and Participation

Who will be your target audience? This will help you determine how to approach the set up of your program. What works with one group will not necessarily work with the next. It is important to be prepared with various teaching methods and be quite adaptable to each audience. Flexibility and adaptability are two hallmarks of an effective teacher. Hands-on activities where, after a subject has been taught, each member of the class practices what they have learned is one of the best ways to incorporate and facilitate participation, and most people will learn through activities where they must participate. This activity needs introduce a new topic or re-enforce one just learned. Interesting and active participatory learning should be an objective of each class.

Four Week Class Program

As an example, you may begin with a diagram similar to the one below. Determine how many weeks you want to teach. Decide what the overarching topic will be and then divide the topic into sub topics, or categories.

Week	Topic	Sub-Topics
Week One	First Trimester and Second Trimester	Conception Fetal development Nutrition (Nutritional Smoothie Activity) Hormones & Emotions Movie *In the Womb*
Week Two	Third Trimester- All About Labor	Signs of Labor Length of Labor Stages of Labor Birth Plan (Activity) Birth Settings Birth Team and Doula
Week Three	Third Trimester Comfort and Coping	Breathing Techniques (Breathing Activity) Comfort Measures Massage (Activity) Labor and Birth Positions Informed Consent
Week Four	Newborn, Postpartum & Breastfeeding	Normal Newborn Breastfeeding Do's and Don'ts Postpartum Expectations Skin to Skin & Moby Wraps (Activity)

Six Week Class Program

Using the example listed above, think about and determine what would be a good teaching approach for a six week course. The main topics are listed for you. Determine what you would teach as sub-topics:

Week	Topic	Sub-Topics
Week One	First Trimester	
Week Two	Second Trimester	
Week Three	Third Trimester	
Week Four	Labor and Birth	
Week Five	Postpartum and Newborn	
Week Six	Breastfeeding	

Eight Week Class Program

An eight week class program might follow the exact same layout and curriculum as the six week course, but it might choose to address special topics. These topics will need to be determined by the audience and what they desire to learn.

Week	Topic	Sub-Topics
Week One	First Trimester	
Week Two	Second Trimester	
Week Three	Third Trimester	
Week Four	VBAC	
Week Five	Labor and Birth	
Week Six	Water Birth	
Week Seven	Newborn and Postpartum	
Week Eight	Breastfeeding	

© GoMidwife 2015

How Will You Market Yourself?

Marketing is difficult for all beginning birth workers and as time progresses your greatest marketing tool will be the word of mouth by happy mothers and families who have previously taken your class. In the beginning, childbirth education may not be a lucrative venture. Most educators do not get into this field for the cash flow, but rather because we believe strongly that there is information we have, that others need. As you develop your course, and offer consistent fruitful classes, then classes can become a reliable source of income. Generally childbirth educators charge in the range of $50-$450 per class. How you set your prices will be determined by several factors. One factor is whether or not you are a beginning or advanced teacher and another is what your community can hold. A good place to begin to market yourself is with local midwives, local doulas, doctors offices, home school co-ops, social media pages, and so forth. There is no wrong place to begin, so be creative in your approach.

Curriculum Development

Writing curriculum may seem daunting, but if taken one topic at a time it will be surprisingly simple. Developing your own curriculum will allow you to compose material based on your philosophy of education and birth and not on someone else. As you take time to write your knowledge will be reenforced, continued learning will likely take place and foundations of understanding will be solidified. Teaching is one of the best ways to learn, and as teachers, we must always be open to continual learning. Here is how to get started:

Writing Your Own Curriculum

Curriculum is simply defined as a course of study. Curriculum is generally designed to help a particular audience achieve specific goals. In the end, your curriculum should be one that not only you are comfortable teaching and that meets a client's needs but also should be written with an eye towards multiplication. If you were to train others, for instance, could you leave a copy with them and have them use it satisfactorily? If it would need a lot of modification you may want to rethink whether or not it is user-friendly.

When writing your curriculum you will need to follow three basic steps:

1. Define the Objective and Determine the Goals

Objectives are the general goals set forth for the class. They are usually broad reaching and state the intended purpose(s) and also the expected outcomes of the class. Objectives can be written for each topic, each trimester or for the class in its entirety. Developing goals and objectives gives you a clear path to navigate in order to reach the desired outcome.

Example:

Goal:

Teach Birth Options

Objective:

Show the movie: *The Business of Being Born*

Discuss: the difference between hospital, birth center and home birth.

Here are some questions to ask as you determine your goals and objectives:

What is the purpose for the course you will create?

What does your audience need to know?

2. Determine Sequential Order and Write an Outline

Create an Outline- Create an outline of key topics to be covered during the class. Determine how many classes are needed to sufficiently cover the material. Identify the sequence in which each topic will need to be taught in order to reach the course objective.

3. Decide the Teaching Approach and Determine the Audience

Often included in the steps of development is an assessment and evaluation process, but for our purposes, we will look at assessment and evaluation separately and a little later.

Determine the Teaching Approach- Lecture, discussion, hand-outs, media and hands-on activities should make up the majority of teaching components. Decide, based on your audience, what the primary approach will be and build the curriculum around it, giving the preference of time to that particular approach. Also, how many classes will be offered in each series? Will it be once a month or once a week? Think about the broad headings: will it be divided by trimesters or by important topics such as nutrition, labor and breastfeeding? These questions will guide you in the flow of the classes and the course as a whole.

Example:

Hands-On Activities- 60%

Discussion- 20%

Media- 15%

Lecture-5%

 **Enrichment:**

Create an evaluation form to be given out at the end of each teaching series.

Creating an Outline

This course was written based on an outline. Go back to the table of contents at the beginning of this manual and read from top to bottom. Prior to this course being written an outline was made and goals and objectives were determined. What was the purpose of this course, what was the desired outcome and what needed to be taught to reach the end goal? A list of topics were determined and the order and sequence in which they needed to be taught in order to build upon themselves was laid out with the result...this course.! Now it is your turn.

Think about those topics that are most important to the area where you will be working. What do they need to know, what are the goals you have in teaching them, and what changes would you like to see occur? These questions should lead you to begin developing your own outline with the things most pertinent to your targeted audience.

 **Enrichment:**

Write an outline for your childbirth education course.

Writing Lesson Plans

1. Determine the main topic for your lesson. An example might be: **Comfort Measures**.

2. Decide the key points you desire to achieve during class. An example might be: positions, massage, hot and cold, movement, water

3. Plan an activity to get the mother and her labor partner involved in the lesson. An example might me: practice labor positions for comfort or provide massage oils and practice effleurage and massage techniques.

4. Establish the amount of time needed for this subject to be taught thoroughly and the time for the activity.

5. Choose supplemental material to support the topic. An example might be: a 5 minute video by Penny Simpkin on Comfort Measures in Labor.

Lesson plans can be as simple or detailed as you choose to make them. The following is an example of a very simple, clear and concise lesson plan.

Topic	Objective	Teaching Method	Time Needed	Activity	Materials
Nutrition	To teach women the importance of nutrition throughout pregnancy	Lecture Discussion	30 minutes 15 mins teaching 15 mins activity	Make nutritional smoothie	chia seed flax seed milk avocado spinach banana apple protein powder

You'll never know everything about anything, especially something you love.
Julia Child

Self Assessment

Evaluation and Feedback- There is always room for improvement. Solicit each mother's help in evaluating the course and how it was taught. This will help know what is working well and what could be improved upon. Have each mother fill out the form autonomously so they can be honest and objective with their answers. Ask open ended questions, for example: What would you do differently in the next class? Ask both for positive and negative feedback such as "what was your favorite part" and "what was your least favorite part". Leave room for comments and suggestions. Hold it loosely. Do not take feedback personally. Be ready to revise and always be flexible.

Sincerely consider how effective your classes are and information is being received. Be willing to adjust to accommodate the need of the mothers. Utilize the feedback received after each class to alter curriculum and the tried approach. Ask hard questions and be impartial in assessing whether or not the goals and objectives set are being reached?

Educators also need to continue to learn. Invite other educators in to your course to participate and evaluate your class. Be open to suggestions and better ways to facilitate information. Join other facilitators classes to see how they are teaching. If in an isolated or rural community take time out to attend a workshop now and again to keep objectivity primed. With each couple you will add an additional piece to enhance your ability to assess the effectiveness of your classes

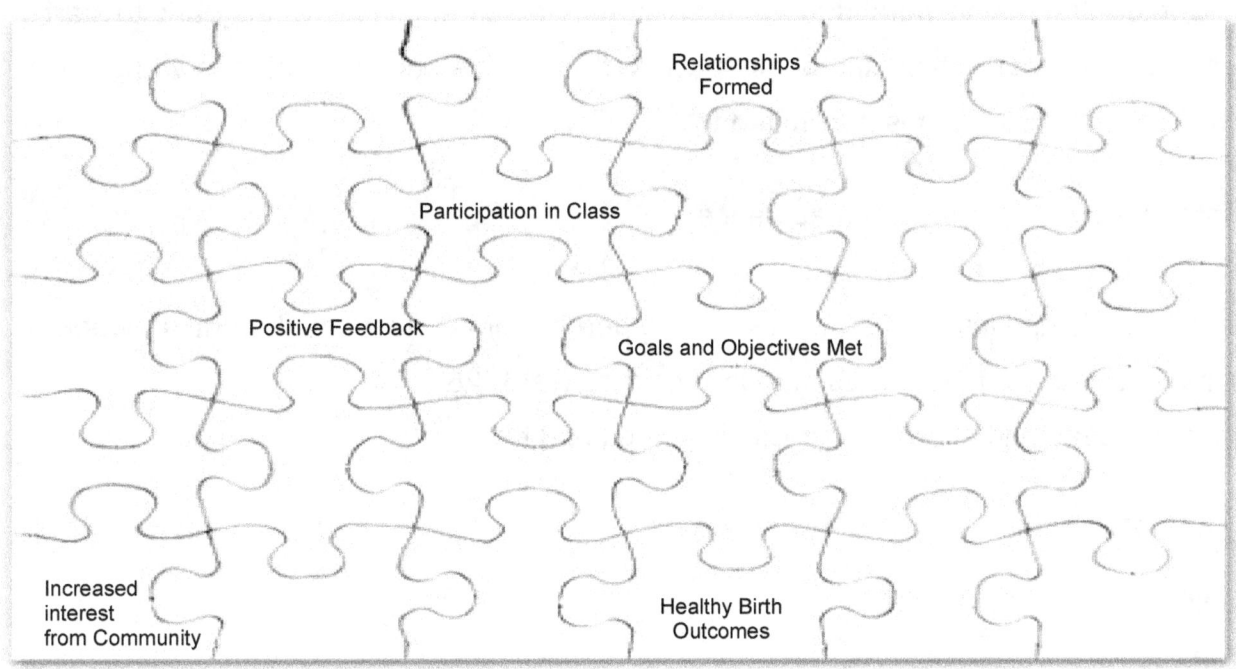

Education is a process, and as you begin your journey as an educator you will not necessarily see the whole picture, nor immediately have all the pieces in the right place. As long as you can add one new piece each time, then you are growing. We all have to begin somewhere. Just as in the picture above, continue to add one piece at a time to the development of your class, until all the pieces are in place to form a holistic and informative childbirth class that truly changes outcomes for mothers, babies, families and communities.

Final Activity

Throughout this course each activity was designed to help you research, think through, create and develop specific topics to be included in your curriculum. If you decide to move towards certification then you should have a childbirth education curriculum ready to use as you begin educating women around the world during the childbearing years. It is never too early to start this process. Gather your outline and begin to work, filling in the subject matter as you go. This curriculum needs to include but is not limited to:

Adapted information to your intended audience

Multifaceted methods of teaching

First trimester needs and information to include the period from conception through 12 weeks

Second trimester needs and information to include weeks 13- 28

Third Trimester needs and information including weeks 28 through 42

Newborn

Postpartum

Breastfeeding needs and information.

Special Topics

Original Handouts (10)

Resourced Hand-Outs (10)

Resource Library: a list of books, movies, videos and online sources

Community Resources: a list of doulas, midwives, naturopaths, pediatricians, general practitioners, OBGYNs, mother's groups, breastfeeding support groups and play groups

You can find excellent resources locally and on the internet to help you on your way. Use them as a guide to make your program look like you! A great resource can be found at the following website:

http://www.transitiontoparenthood.com/ttp/foreducators/outlines/outlinehome.htm

Education is the key to unlock the golden door of freedom.
George Washington Carver

END NOTES

Section

1. **Why Become a Childbirth Educator**, p. 10

The World Health Organization, http://www.childbirthconnection.org/article.asp?ck=10456

2. **Why Become a Childbirth Educator**, p. 10

The World Health Organization, http://www.who.int/mediacentre/factsheets/fs348/en/

3. **Performance Based Birth**, p. 25

Information from   http://www.thesaurus.com/

4. **Effects of Childbirth Education**, p. 27

Koehn, Mary L. *The Journal of Perinatal Education.* 2002 Summer; 11(3): 10–19.

5. **Learning Styles**, p. 33

Definition from
http://blc.uc.iupui.edu/AcademicEnrichment/StudySkills/LearningStyles/3LearningStyles.aspx

6. **Learning Styles**, p. 33

Definition from http://www.learning-styles-online.com/overview/

7. **Teaching Survivors**, p. 37

http://www.pandys.org/articles/survivorsgivingbirth.html

8. **Understanding Conception**, p.40

Image from Health Info, 2011, http://health-online-info.blogspot.com/2012/02/masculine-mens-sperm-fewer.html

9. **Physiology in Pregnancy**, p. 43

© GoMidwife 2015

Information from: http://almostadoctor.co.uk/content/systems/obstetrics-and-gynaecology/pregnancy-and-labour/normal-physiology-pregnancy

as well as

http://www.coursewareobjects.com/objects/evolve/E2/book_pages/lowdermilk/pdfs/208-230_CH08_Lowdermilk.qxd.pdf

10. **Aromatherapy**, p. 62

Information from http://articles.mercola.com/sites/articles/archive/2014/09/04/essential-oils-aromatherapy.aspx

and

http://www.herbsofgrace.com/Blog/2004/01/25/aromatherapy-for-pregnancy-by-vanessa-nixon-klein-owner-herbs-of-grace-inc/

11. **VBAC – Vaginal Birth After Cesarean**, p. 66

ACOG information taken from http://www.vbac.com/acogs-revised-vbac-guidelines/

12. **Cesarean**, p. 69

Quote taken from http://www.childbirthconnection.org/pdfs/cesarean-section-trends.pdf

13. **Birthing Techniques and Methods**, p. 70

Definition from www.businessdictionary.com

14. **Weight in Pregnancy**, p. 72

Information from http://www.askdrsears.com/topics/pregnancy-childbirth/pregnancy-concerns/gaining-weight/healthy-weight-gain-during-pregnancy

15. **Get a Doula**, p. 73

Quote from http://www.ncbi.nlm.nih.gov/pmc/articles/PMC3647727/

16. **Fetal Kick Counts**, p. 75

Information from   http://americanpregnancy.org/duringpregnancy/kickcounts.htm

17. **Fetal Proteins, Development and the Indication of Labor**, p. 88

Article from

Lu Gao, Elizabeth H. Rabbitt, Jennifer C. Condon, Nora E. Renthal, John M. Johnston, Matthew A. Mitsche, Pierre Chambon, Jianming Xu, Bert W. O'Malley, Carole R. Mendelson. **Steroid receptor coactivators 1 and 2 mediate fetal-to-maternal signaling that initiates parturition**. Journal of Clinical Investigation, 2015; DOI: 10.1172/JCI78544

and from

http://www.sciencedaily.com/releases/2015/06/150622162023.htm

18. **Birth Rights**, p. 89

Information from   http://gomidwife.com/induction-of-labor-and-rights/

19. **Normal Newborn Development**, p. 107

Growth Spurt information from   http://kellymom.com/bf/normal/growth-spurts/

www.ingramcontent.com/pod-product-compliance
Lightning Source LLC
Chambersburg PA
CBHW080920170526
45158CB00008B/2173